This book is to be returned on or before
the last date stamped below.

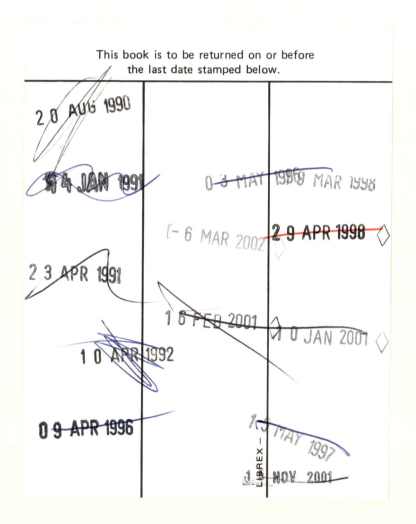

2 0 AUG 1990

4 JAN 1991

0 3 MAY 1996   9 MAR 1998

1 - 6 MAR 2002   2 9 APR 1998

2 3 APR 1991

1 6 FEB 2001   1 0 JAN 2001

1 0 APR 1992

0 9 APR 1996

1 3 MAY 1997

LIBREX —

1 NOV 2001

Engine Electronics

Today's traffic puts great demands on an automobile. These demands must be mastered as well as possible even under extreme conditions. Automotive electronics plays a decisive role in this task in the areas of ignition and fuel management.

The economical or improved use of fuel in conjunction with low-pollutant combustion can also be realised with the help of electronics.

In particular, microelectronics is the technology used to solve control problems safely, reliably and economically.

All of this is possible only with sensors which provide the operating data for the electronics. For example, sensors are used to measure temperature, pressure and engine speed with maximum precision. This is a definite must if engine electronics are to succeed.

Prepared in cooperation with the appropriate technical departments of the Robert Bosch GmbH.

A number of illustrations were kindly provided by: Kolbenschmidt AG, Neckarsulm and Volkswagen AG, Wolfsburg.

Electronic Ignition Triggering

The Problem

The ignition contact of a coil ignition, and also the control contact of a breaker-triggered transistorized ignition, were subjected to mechanical wear because of the on/off intermittent operation. The ignition point did not remain as constant as desired over the operational life so that maintenance became necessary.

The Solution

The control or ignition contact is replaced with non-contacting (= non-wearing) electronic sensors which drive the switching transistors in the trigger box.

The Advantages

● Non-wearing.
● Maintenance-free.
● Uniform ignition signals.
● Constant ignition point.

Inductive Ignition Triggering

Induction-Type Pulse Generator in the Ignition Distributor
(Figures 1, 2 and 3):
The permanent magnet (1) and induction winding (2) form the stator. The timer core which is mounted on the distributor shaft forms the rotor which rotates past the stator. The core and the rotor are made of soft magnetic steel. They each have protruding teeth (stator teeth and rotor teeth). As the rotor turns, the air gap (3) between the stator and rotor teeth changes periodically. The magnetic flux changes as the air gap changes. This change in the flux then induces an alternating voltage in the induction winding.

Origin of the Generator Voltage
As the rotor teeth approach the stator teeth, the magnetic flux intensifies. This change in the flux induces a voltage in the induction winding. This voltage increases up to a maximum value which is reached shortly before the teeth are directly opposite each other. As the timer core continues to rotate, the distance between the teeth increases and the generator voltage changes its sense. The pulse generator is used to generate an alternating voltage which is used, instead of contacts, for controlling the ignition. The frequency of this alternating voltage corresponds to the spark frequency.

Induction-type Pulse Generator Mounted on the Crankshaft
(Figures 4, 5 and 6)
Depending upon the requirements placed on triggering accuracy and the engine design, various triggering systems on the crankshaft are possible:
1. An inductive speed sensor and an inductive reference-mark sensor. These devices sense the teeth of the ring gear on the flywheel or a pin mounted on the flywheel without making contact.
2. An inductive sensor which senses a special toothed disc on the crankshaft and supplies a signal. This signal is used jointly to determine the engine speed and the crankshaft position.
3. An inductive sensor which is affected by a special segmented disc mounted on the crankshaft. This sensor supplies not only a signal for the engine speed but also one for the crankshaft position.
In this manner, both the engine speed and the crankshaft position can be determined.
The values determined at the crankshaft are more accurate than those measured using the ignition distributor.

D
624.254
ENG

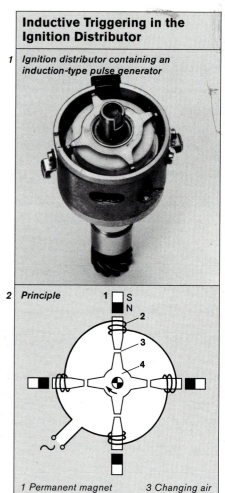

Inductive Triggering in the Ignition Distributor

1 *Ignition distributor containing an induction-type pulse generator*

2 *Principle*

1 Permanent magnet
2 Induction winding with core
3 Changing air gap
4 Timer core

3 *Curve of the induction voltage versus time*

Voltage
0

t_Z t_Z
Time ⟶

M.F.F

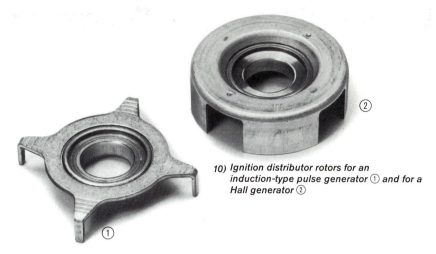

10) *Ignition distributor rotors for an induction-type pulse generator* ① *and for a Hall generator* ②

Inductive Triggering at the Crankshaft

4   *Toothed disc (on the crankshaft) with an induction-type pulse generator*

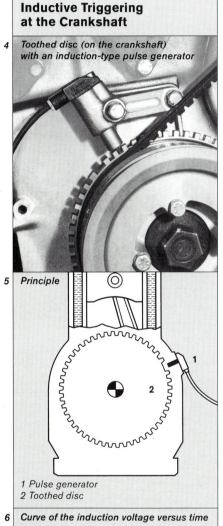

5   *Principle*

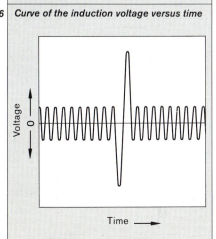

1 Pulse generator
2 Toothed disc

6   *Curve of the induction voltage versus time*

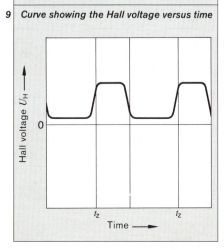

Voltage

Time ⟶

Triggering by means of a Hall Gen. in the Ignition Distributor

7   *Ignition distributor containing a Hall generator*

8   *Principle*

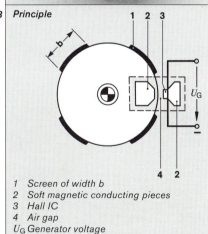

1   Screen of width b
2   Soft magnetic conducting pieces
3   Hall IC
4   Air gap
U_G Generator voltage

9   *Curve showing the Hall voltage versus time*

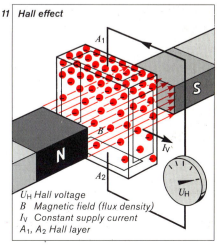

Hall voltage U_H

t_z t_z

Time ⟶

Ignition Triggering by means of a Hall Generator

The Hall Generator

(Figures 7, 8 and 9)

If the ignition distributor shaft rotates, the screens of the rotor pass through the air gap of the magnetic barrier without making contact. If the air gap is open, the IC and, with it the Hall layer, is permeated by the magnetic field.

If the magnetic flux density is high at the Hall layer, the Hall voltage U_H is at its maximum. The Hall IC is active. As soon as one of the screens enters the air gap, the magnetic flux is absorbed to a great degree by the screen and thus does not reach the IC. The flux density at the Hall layer is reduced to a small residual density which is caused by the stray field. The voltage U_H is now at its minimum.

The Hall Effect

If electrons pass through a conductor which is permeated by the lines of force of a magnetic field, the electrons are deflected perpendicular to the direction of current flow and perpendicular to the direction of the magnetic field: An excess number of electrons result at A_1 while the area around A_2 is depleted of electrons, i.e. the Hall voltage exists between A_1 and A_2. This so-called Hall effect is particularly pronounced in semiconductors (Fig. 11).

11   *Hall effect*

U_H Hall voltage
B Magnetic field (flux density)
I_V Constant supply current
A_1, A_2 Hall layer

Electronic Switching of the Primary Current

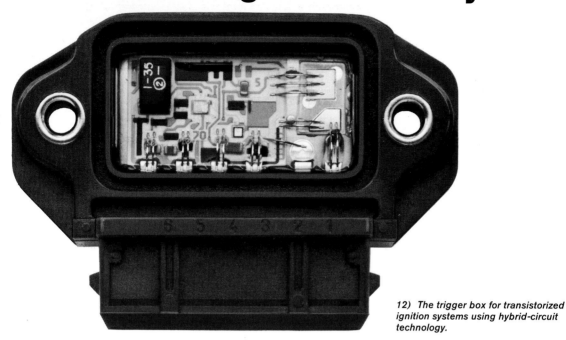

12) *The trigger box for transistorized ignition systems using hybrid-circuit technology.*

The Problem

To ensure the ignition of the air-fuel mixture, it is necessary to provide an adequate ignition voltage and a specific amount of ignition-spark energy for the spark plug.

The limited capacity of the contact-breaker points meant that it was not possible to increase the primary current (above 4.5 A). This was necessary to obtain increased spark energy and increased secondary voltage, particularly at high engine speeds.

The Solution

Switching transistors were used to carry the primary current. These transistors were originally controlled by control contacts and then by electronic triggering systems.

The Advantages

● Zero-loss and non-wearing switching of high primary currents (up to approx. 9 A) resulting in a high secondary voltage over the entire engine speed range.
● Maintenance-free triggering systems.

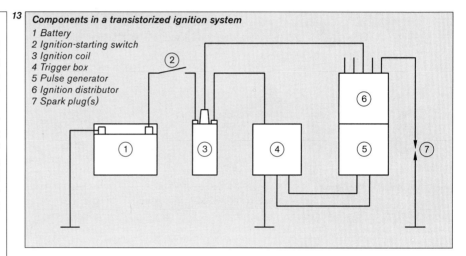

Components in a transistorized ignition system
1 Battery
2 Ignition-starting switch
3 Ignition coil
4 Trigger box
5 Pulse generator
6 Ignition distributor
7 Spark plug(s)

The System

The ignition system consists of the following components:
① The battery as the energy source. The primary current flows from the battery to the ② ignition-starting switch (when it is switched on), through the primary winding of the ③ ignition coil, via the ④ trigger box and then back to the battery. Interruption of the primary current by the trigger box is controlled by a ⑤ pulse generator in the ⑥ ignition distributor.

In older systems, this pulse generator may be a mechanical control contact. In more recent systems, this task is performed by an electronic ignition-triggering device.

At that instant when the primary current is interrupted, high voltage is induced in the secondary winding of the ignition coil. This high voltage is distributed by the ignition distributor to the appropriate ⑦ spark plug.

The Trigger Box

The following description uses a transistorized ignition system with an induction-type pulse generator as an example.

Design

(A trigger box for a 6-cylinder engine) The components are mounted on a printed board and electrically connected to one another by means of the conductors printed onto the back of the board. The printed board is permanently mounted on a metal base. This base conducts the waste heat generated by the circuit to the mounting surface. A plastic cap protects the components against dirt and mechanical damage. The connector with blade contacts is located on the side of the cap.

Operation

The alternating control voltage coming from the induction-type pulse generator must be reshaped into rectangular current pulses in order to have the correct effect in the trigger box. Thus reshaping is done by an electronic threshold switch which is known in control engineering as a Schmitt trigger. This circuit is termed a pulse-shaping circuit because of its function in the trigger box. The components of section B shown in Figure 14 are part of the pulse-shaping circuit. D4 is a silicon diode; due to its polarity, it allows only the negative pulses of the alternating control voltage to reach the base B of the transistor T1 whereas the positive pulses are blocked.*) The induction-type pulse generator is electrically loaded only in the negative phase of the alternating control voltage because of the output of energy. In the positive phase, on the other hand, the pulse generator is not loaded. For this reason, the negative voltage amplitude (shown at the left of Figure 15) is smaller than the positive amplitude.

As soon as the alternating control voltage, approaching from negative values, exceeds a threshold at the input to the pulse-shaping circuit, transistor T1 (Figure 14) conducts and T2 now switches to the blocking state and does not allow current to pass. The output of the pulse-shaping circuit does not carry current for the time T_i. This switching state is maintained until the alternating control voltage, now approaching from positive values, drops below the threshold voltage.

Transistor T1 now blocks for time T_0. The base B of T2 becomes positive via R5 and causes T2 to conduct. This alternation – T1 conducts/T2 blocks or T1 blocks/T2 conducts – is typical of the Schmitt trigger and the circuit repeats this action continuously. The two series-connected diodes D2 and D3 are provided for temperature compensation.

The energy stored in the ignition coil can be put to optimum use with the help of Section C in the trigger box. The result is that sufficient high voltage is available for the spark at the spark plug under any operating condition of the engine. This so-called dwell-angle control specifies the start of the "dwell period" T_i. The term "dwell period" (= dwell angle) is borrowed from breaker-triggered ignition systems and signifies the time from when contact is first made until contact is broken. The "beginning of the dwell period" is also the beginning of a rectangular current pulse which is used to trigger the "driver" in the trigger box (transistor T4). A timing circuit using RC elements is used here. This circuit alternately charges and discharges capacitors by way of resistors.

The current from the driver transistor T4 drives the power output stage (a Darlington circuit, Figure 14). In this Darlington circuit, the current flowing into the base B of transistor T5 is amplified to a considerably higher current which is fed into the base of the transistor T6. Then, the high primary current is fed to the ignition coil through T6. The primary current is switched as the collector current of this transistor. The Darlington circuit functions as one transistor having the terminals B, C and E.

In addition to the device described above, there is also a trigger box for Hall generators. This type of trigger box performs basically the same functions as those described above. As early as 1978, Bosch introduced the first hybrid ignition trigger boxes to the market. These hybrid ignition trigger boxes are considerably smaller than those utilizing discrete components and they are available both for ignition systems with induction-type pulse generator drive as well as those with Hall generator drive.

14

Circuit diagram of a trigger box for transistorized ignition systems (6-cylinder engine).

A *Voltage-regulation components*
B *Pulse-shaping circuit components*
C *Dwell-angle control components*
D *Darlington output stage (IC)*

    Primary current
    Path of the control pulse

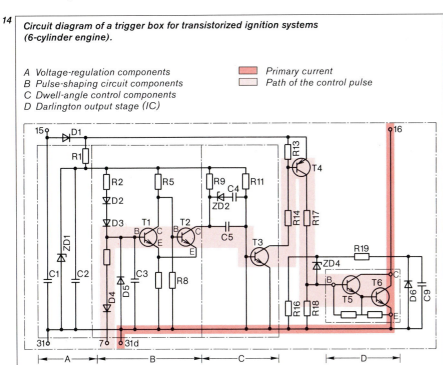

15

Pulse diagram of a transistorized ignition system with an induction-type pulse generator.

The figure illustrates the entire pulse processing system of a breakerless transistorized ignition system. It starts at the pulse generation by the induction-type pulse generator and proceeds up to the spark discharge at the spark plug. As shown in this pulse diagram, the alternating control voltage from the induction-type pulse generator is fed to the pulse-shaping circuit which converts the voltage into rectangular current pulses. The pulse duration (this corresponds to the dwell angle) is increased or decreased by the control system (the dwell-angle control, described in a following section) as a function of the engine speed.

The current of the rectangular pulses is amplified in the driver and is then used to drive the output transistor. The output transistor switches the primary coil current (red) on and off in time with the pulses. Each interruption of the rectangular pulses interrupts the primary current, also interrupting the spark discharge at the spark plug at the ignition point t_z.

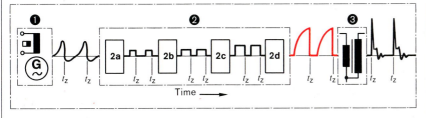

t_z *Ignition point*
❶ *Induction-type pulse generator*
❷ *Trigger box*
2a *Pulse-shaping circuit*

2b *Control for the pulse duration (dwell angle)*
2c *Driver*
2d *Darlington output stage*
❸ *Ignition coil*

*) "Positive" and "negative" refer to the potential at terminal 7 with respect to the ground potential at terminal 31 or 31d.

Electronic Dwell-Angle Control

The Problem

The time between the control signals of the trigger system differs as a function of the engine speed. However, a specific minimum primary current is necessary to achieve a constant ignition energy. To achieve this minimum primary current, in turn, a specific dwell period is necessary. At high engine speeds, this required dwell period is not always reached. This may result in misfiring in the upper speed range.

The Solution

An electronic circuit controls the cut-in time for the primary current, as a function of the engine speed and the battery voltage, such that the minimum current is reached by the time the ignition point occurs.

The Advantages

The dwell period matched to the operating condition and the resulting primary current ensures high ignition energy, avoids misfiring in all ranges of operation and thus saves fuel.

The System

In order to maintain the ignition power of the system at a constantly high level, while at the same time guaranteeing minimum power loss in output transistor and ignition coil, the primary current at the moment of ignition point must have reached a given level.
The time that current flows through the ignition coil is controlled as a function of the engine speed and battery voltage such that, in stationary operation, the desired primary current has just been achieved at the end of the time current is flowing through the ignition coil. A dynamic correction is superimposed at low engine speeds so that, during acceleration, the minimum primary current is achieved because of this extra current in spite of the reduced dwell period. The output stage acts as a current limiter so that if the desired primary current is achieved before the ignition point, the primary current is held constant until the ignition point.
The dwell angle is equal to the angle of rotation of the distributor cam on the distributor shaft from the time contact

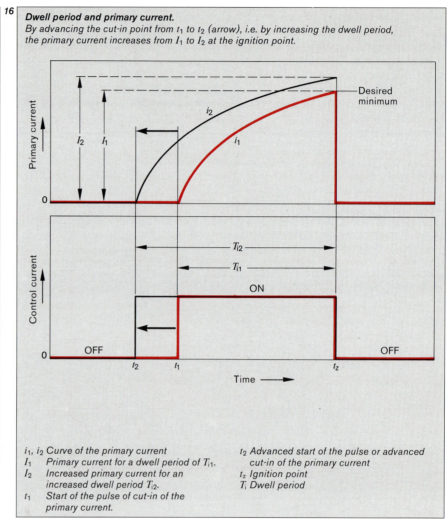

16

Dwell period and primary current.
By advancing the cut-in point from t_1 to t_2 (arrow), i.e. by increasing the dwell period, the primary current increases from I_1 to I_2 at the ignition point.

i_1, i_2 Curve of the primary current
I_1 Primary current for a dwell period of T_{i1}.
I_2 Increased primary current for an increased dwell period T_{i2}.
t_1 Start of the pulse of cut-in of the primary current.
t_2 Advanced start of the pulse or advanced cut-in of the primary current.
t_z Ignition point
T_i Dwell period

is made in the contact breaker until contact is broken. The time which passes during this process is known as the dwell period. The quantities of dwell angle and dwell period are proportional to one another. The dwell period is equal to the time during which the primary current is flowing. The energy stored in the ignition coil can be put to optimum use with the help of the dwell-angle control so that sufficient high voltage is available for the spark discharge under any operating condition of the engine. In order to provide voltage which is independent of the engine speed, the dwell-angle control increases the dwell angle as the engine speed increases. The dwell angle is also increased in order to be able to achieve the desired primary current even if the battery voltage drops.
The dwell angle is limited to a maximum value (Figure 17) in order to guarantee a minimum time during which contact is broken and thus a minimum

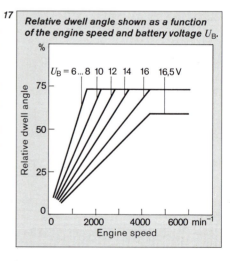

17

Relative dwell angle shown as a function of the engine speed and battery voltage U_B.

spark duration. To achieve this, the dwell-angle control operates so that, depending on the engine speed and the battery voltage, the dwell angle is changed by varying the start of the dwell period whereas the end of the dwell period and thus the ignition point are not affected.

18

Operation of the simple dwell-angle control with two switching states:

a) Capacitor charging (charging phase)

b) Capacitor discharging (control phase)

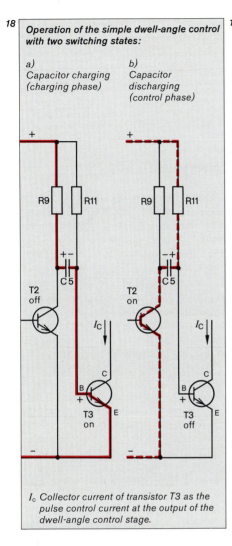

I_c Collector current of transistor T3 as the pulse control current at the output of the dwell-angle control stage.

19

Pulse train of the dwell-angle control and curve of the voltage on capacitor C5 shown versus time. The signs (−) and (+) refer to the potential at the capacitor connection shown at the right in Figure 18b.

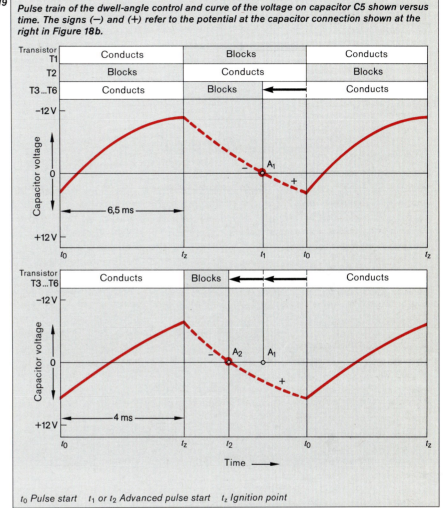

t_0 Pulse start t_1 or t_2 Advanced pulse start t_z Ignition point

Open-loop Control

The electronic dwell-angle control is a timing control using RC elements. It alernately charges and discharges capacitors via resistors. This is an open-loop dwell-angle control because the combination of resistors and a capacitor provides a fixed timing relationship as a function of the engine speed (for a closed-loop dwell-angle control, see Page 8). It is an analog control system because the dwell period can change continuously (within certain limits). The switching stage of a simplified open-loop dwell-angle control system is shown in Figure 18. The capacitor C5 and the two resistors R9 and R11 serve as the RC element. When the trigger transistor T2 is in the blocking state, the capacitor C5 is charged via R9 and the base-emitter path through transistor T3 (18a, red). At a low engine speed, the capacitor voltage reaches almost 12 V. During this time T3 conducts

(Figure 18a). At the ignition point t_z, T2 conducts and C5 can now discharge via R11 and T2. As long as the discharge current is flowing T3 is off because its base B is negative with respect to ground. T3 starts to conduct again at that instant when the polarity of the voltage across capacitor C5 has changed. Thus, when positive becomes negative and vice versa*). The capacitor now is charged in the opposite sense (Figure 18b) until transistor T2 enters the blocking state. At this point, the charge on the capacitor is reversed by way of R9 and the base-emitter path through T3 starting at time t_1 (Figures 18a and 19, red).

The functional characteristic of the open-loop dwell-angle control is the cut-in of the collector current of transistor T3 when the polarity changes (Fig. 19, top, point A_1). The other transistors T4 to T6 also start to conduct at the same time as T3 (Figure 14, Page 5, the section entitled "Electronic Primary-current Switching"). The desired

advance of the beginning of the dwell period (Figure 19, arrow) referenced to the cut-in time t_0 of the trigger transistor T_1 is achieved in this manner. As the engine speed increases, the charging time of the capacitor C5 decreases. The capacitor can no longer be charged to the full battery voltage. C5 is discharged correspondingly earlier resulting in an even earlier start of the dwell period t_2 and thus an even larger relative dwell angle (Figure 19 bottom, point A_2).

*) To be more precise: T3 starts to conduct at that instant when the charge across the capacitor shown in Figure 18b has reversed to such an extent that the threshold voltage is present between the base B and the emitter E of T3.

Electronic Closed-loop Dwell-angle Control, Primary-current Limiting and Stationary-engine Cut-off

The Problem

The combustion process in the cylinder can be improved even more by making high ignition energy available. In this manner, it becomes possible to provide an improved engine design.

The Solution

If a closed-loop dwell-angle control system with current limiting is super-imposed on the open-loop dwell-angle control, it can be ensured that the correct primary current is achieved at the ignition point under all operating conditions.

Switching off the primary current when the engine is stationary prevents excessive heating of the coil when the ignition is switched on. Both measures make it possible to provide an improved layout of the coil and the output transistor (of the transistorized ignition system) to provide high spark energy with a low power loss.

The Advantages

● With **current limiting**, ballast resistors and the associated cables and mounting hardware as well as terminal 15a in the starter-motor relay (voltage increase for starting) are no longer necessary.
● With **closed-loop dwell-angle control**, the dependence of the ignition system on the battery voltage, temperature and engine speed is reduced.
● The **stationary-engine cut-off** prevents the primary current from flowing when the ignition is switched on and the engine is stationary.
● Overall, the availability of high-voltage and energy for the ignition process is improved thus providing increased reliability for combustion of the air-fuel mixture.

Design

(The trigger box uses hybrid technology)

The plastic trigger-box housing, together with the connector and the injection-molded blade terminals form a single unit. A metallic base plate is used as the support for the circuit while also dissipating the heat. In order to prevent excessive heating of the hybrid circuit, the power output stage is not located on the ceramic substrate but rather it is insulated and mounted on the metal base plate. The base plate is bonded to the plastic housing by means of adhesive.

To protect the circuit against moisture, the interior of the trigger box is filled with silica gel and its cover is attached by means of adhesive. The trigger box is mounted to the vehicle body together with a special heat sink.

Operation

Closed-loop Dwell-angle Control

The dwell-angle is controlled as far as is technically feasible so that the same primary cut-off current is achieved in any operating condition, that is, with differing battery voltage, engine speed and temperature. This is shown clearly in Figure 23. For example, with a battery voltage of 6 V (during starting) the curve of the primary current is flatter than it is for 12 V or 15 V. This means that, in the case of 6 V, the primary current must be switched on considerably earlier – thus the dwell angle must be increased – than is the case for 12 V or 15 V. In order to keep the average power loss and thus the heating of the ignition system low, the dwell angle is regulated so exactly that the current is limited for only a short time, in terms of per cent, between ignition points. In an ignition distributor equipped with an **induction-type pulse generator**, the dwell angle is changed by shifting

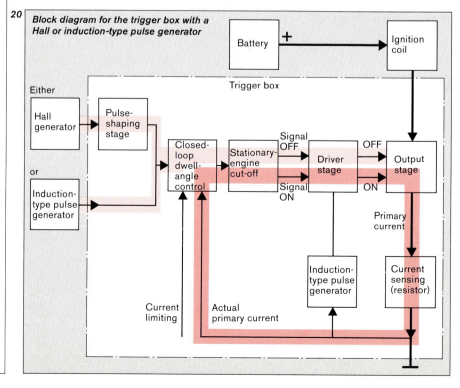

20　*Block diagram for the trigger box with a Hall or induction-type pulse generator*

Battery　+　Ignition coil

Trigger box

Either

Hall generator → Pulse-shaping stage

or

Induction-type pulse generator

→ Closed-loop dwell-angle control → Stationary-engine cut-off → Signal OFF / Signal ON → Driver stage → OFF / ON → Output stage

Induction-type pulse generator

Current sensing (resistor)

Primary current

Current limiting　　Actual primary current

the trigger level in the trigger box. The cut-in and cut-out points of the trigger move along the curve of the pulse-generator voltage. If the dwell angle is too small, the trigger level is shifted further in the negative direction. In the case of a large dwell angle, the process is reversed (Figure 21).

In an ignition distributor equipped with a **Hall generator**, a pulse-shaping stage must precede the trigger since the Hall generator does not provide an analog signal such as that provided by the induction-type pulse generator. The pulse-shaping stage converts the rectangular signal supplied by the Hall generator into a triangular ramp voltage or sawtooth voltage. In the event of a change in the dwell angle, the trigger levels move along this ramp voltage (Figure 22).

Primary-current Limiting

Due to the fact that ballast resistors are no longer in the circuit, the output stage – in contrast to older transistorized ignition systems – must also perform current limiting. Because of this, ignition coils with a low-resistance primary winding may be used. The maximum primary current is now no longer specified by the total resistance of the primary circuit but rather by the current limiting action in the trigger box. The desired primary current is specified by setting the current limit in the trigger box. Current limiting then functions (in a simplified sense!) such that when the desired primary current is reached at the current-sensing resistor (Block diagram 20), a defined voltage drop is created. This voltage drop is recognized by the current limiting circuit and the circuit causes the output transistor to operate like an electronically controlled ballast resistor. The voltage dropped across the output transistor can thus take on different values: With the current-limiting output stage, approx. 6 to 8 V are dropped across the conducting output transistor during the actual current-limiting period.

Stationary-engine Cut-off

So that the ignition system is not overloaded when the engine is stationary and the ignition is switched on, the transistorized output stage is switched off electronically after one second at the most. As soon as the engine is started, the system outputs ignition sparks again immediately.

Driver stage

The driver stage with voltage limiting and interference suppression corresponds to that used in conventional transistorized ignition systems.

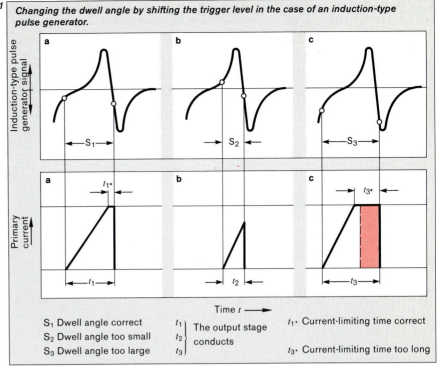

21 Changing the dwell angle by shifting the trigger level in the case of an induction-type pulse generator.

Induction-type pulse generator signal — a, b, c

Primary current — a, b, c

Time t —>

S_1 Dwell angle correct
S_2 Dwell angle too small
S_3 Dwell angle too large

$\left.\begin{array}{c}t_1\\t_2\\t_3\end{array}\right\}$ The output stage conducts

t_{1*} Current-limiting time correct

t_{3*} Current-limiting time too long

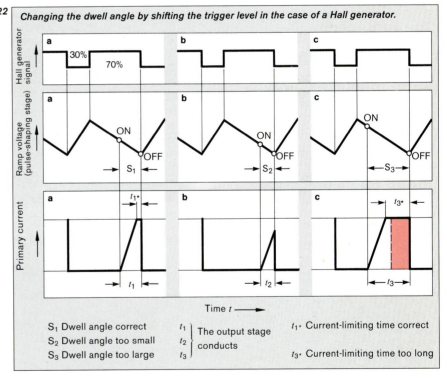

22 Changing the dwell angle by shifting the trigger level in the case of a Hall generator.

Hall generator signal — a (30% / 70%), b, c

Ramp voltage (pulse-shaping stage) — a (ON / OFF S_1), b (ON / OFF S_2), c (ON / OFF S_3)

Primary current — a (t_{1*}, t_1), b (t_2), c (t_{3*}, t_3)

Time t —>

S_1 Dwell angle correct
S_2 Dwell angle too small
S_3 Dwell angle too large

$\left.\begin{array}{c}t_1\\t_2\\t_3\end{array}\right\}$ The output stage conducts

t_{1*} Current-limiting time correct

t_{3*} Current-limiting time too long

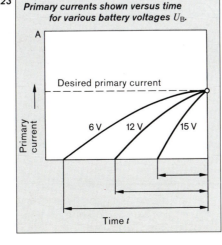

23 Primary currents shown versus time for various battery voltages U_B.

A

Desired primary current

6 V   12 V   15 V

Primary current

Time t

21) and 22)

a The primary current reaches the desired current. The current-limiting time is not too long.

b The primary current does not reach the desired current, e.g. in the case of high acceleration. By means of the closed-loop dwell-angle control, the dwell angle is increased in the next cycle to such an extent that the desired primary current can be reached again.

c The primary current reaches the desired current but the current-limiting time is too long, e.g. due to heavy engine deceleration. In the next cycle, the dwell angle is reduced by the red amount. Note: The red area is only converted to heat.

Electronic Spark Advance

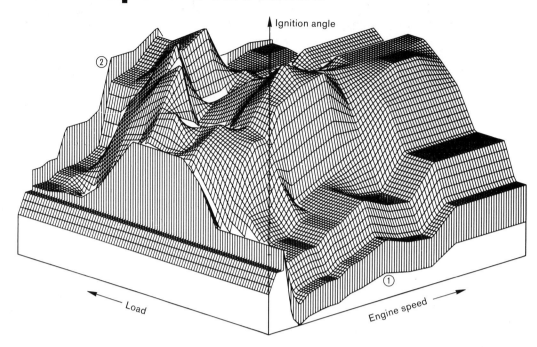

The Problem

Transistorized ignition systems with a conventional distributor with centrifugal and vacuum advance can only achieve simple spark advance characteristics (Fig. 25). This does not fully meet the requirements of engines.

The Solution

Mechanical spark advance in the ignition distributor is no longer used. The pulse-generator signal, which is already available for ignition-triggering purposes, is used in its place as the engine-speed signal. An additional pressure sensor supplies the load signal. The microcomputer calculates the required ignition timing advance and appropriately modifies the output signal which is fed to the trigger box.

The Advantages

● The ignition advance can be better matched to the individual and various demands which are placed on the engine.
● It becomes possible to include additional control parameters (e.g. engine temperature).
● Good starting behavior is provided, idle-speed control is improved and fuel consumption reduced.
● The sensing of operating data can be expanded.
● Knock control can be implemented.

Operation

Figure 24 depicts an ignition characteristic map such as those which can be implemented for electronic ignition advance using a microcomputer. In contrast, Figure 25 shows the ignition characteristic map for a mechanical and vacuum-controlled system.

The signal output by a vacuum sensor is used as the load signal for the ignition system. This signal and the engine speed are used to construct a three dimensional ignition characteristic map. This map makes it possible to program the best, in terms of emissions and fuel consumption, ignition angle (along the vertical) for each speed and load point (the horizontal plane).

The entire characteristic map contains a total of approx. 1000 ... 4000 separately accessible ignition angles depending upon the requirement.

With the throttle valve closed, the lowest line of the characteristic map is selected as the idle/overrun characteristic (Figure 24 ①). For speeds below the desired idle speed, the ignition angle is advanced in order to stabilize the idle speed by increasing the torque. On the overrun the spark advance is programmed for optimum exhaust and driveability.

At full load, the top line of the ignition characteristic map is selected (Figure 24 ②). At this point, the ignition value most favorable with respect to the engine knocking limit is programmed. For starting, an ignition-angle curve independent of the ignition characteristic map is programmed as a function of the engine speed and engine temperature. In this manner, a high engine

24) Ignition characteristic map for the electronic spark advance.
When compared to Figure 25 (the characteristic map of a mechanical ignition advance system), it can be seen that every operating point in the load-speed plane is assigned its own ignition angle with the help of the electronic ignition advance.

torque can be achieved during starting without encountering any restoring torque.

Depending upon the requirements, characteristic maps of various complexity can be implemented or it is also possible to provide only a few programmable ignition advance curves. Electronic ignition advance can be incorporated in various electronic ignition systems. For example, a fully integrated ignition advance system is provided in Motronic.

However, an electronic ignition advance can also be implemented as an addition to a transistorized ignition system (in the form of an added spark-advance mechanism).

25

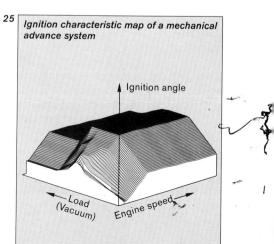

Ignition characteristic map of a mechanical advance system

26) *Electronic spark advance unit with aneroid box (D) for measuring the intake manifold pressure.*

D

Spark Advance Unit

The circuit for electronically advancing the ignition point can be implemented by using a suitable microcomputer.

The input quantities, such as digital switch or pulse signals and analog sensor or pulse signals, are fed to the microcomputer either directly or by way of the analog-digital converter (Figure 27).

Input signals

The two main quantities for controlling the ignition point are the engine speed/crankshaft position and the intake manifold pressure.

Engine Speed and Crankshaft Position: An induction-type pulse generator which senses the teeth of a special toothed wheel mounted on the crankshaft is used to measure the engine speed. An alternating voltage is induced by the change in the magnetic flux produced by the generator. This alternating voltage is evaluated by the control unit.

To precisely measure the crankshaft position, this toothed wheel has a gap which is sensed by the induction-type pulse generator. This signal is then processed in a special circuit. It is also possible to sense the engine speed and the reference mark using a Hall generator in the ignition distributor.

Load (Intake Manifold Pressure): The pressure in the intake manifold acts on the diaphragm of the pressure sensor by way of a tube. The deflection of the diaphragm shifts the position of a magnet. This results in the change of a magnetic field which acts on a semiconductor element. The Hall voltage output by this semiconductor element is directly proportional to the magnetic field and thus to the intake manifold pressure.

Engine Temperature: A water-temperature sensor mounted in the engine block provides the control unit with a signal corresponding to the engine temperature.

Battery Voltage: This is also a correction quantity measured by the control unit.

Throttle-valve Position: A throttle-valve switch supplies a switching signal when the engine is at idle and at full-load.

Signal Processing: The intake manifold pressure, engine temperature and battery voltage, as analog quantities, are digitized in the analog-digital converter. The engine speed, crankshaft position and throttle-valve position are digital quantities and are fed directly to the microcomputer.

The microcomputer processes the signals. The microcomputer consists of the microprocessor with a quartz oscillator crystal for generating the clock signals and a progammable read-only memory with a random access memory for rapidly changing data. Values for the ignition and dwell period are updated here up to 9300 times a minute. These values are recalculated in order to be able to provide the optimum ignition point – this is the output quantity – for the engine at any operating point.

Ignition Output Signal

The primary circuit of the ignition coil is switched by a power output stage in the electronic control unit. The dwell period is controlled such that the secondary voltage remains virtually constant regardless of the engine speed and battery voltage.

The secondary voltage reaches the spark plugs from the ignition coil by way of the high-tension distributor. Since the electronic control unit provides for ignition advance and triggering of the ignition process the only task of the high-tension distributor is to distribute the high tension when the speed and reference mark are sensed on the crankshaft.

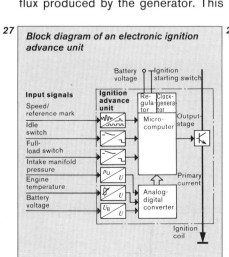

27 | *Block diagram of an electronic ignition advance unit*

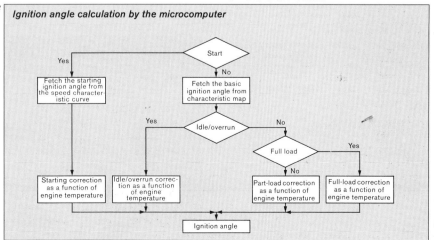

28 | *Ignition angle calculation by the microcomputer*

Electronic Knocking Control

(Foto: Kolbenschmidt)

29) Piston damage due to knocking
Following continuous knocking combustion, the piston looks as if it has been damaged by a welding torch.

The Problem

A high compression ratio is desirable in order to reduce the consumption and increase the torque. However, as the compression ratio increases, so does the danger of an uncontrolled spontaneous ignition of the air-fuel mixture. This results in knocking combustion.

The Solution

The vibrations emanating from the combustion chamber are measured by a sensor, detected by an evaluation circuit and fed to the control circuit. In this case, the control unit retards the ignition point until the engine drops below the knocking limit again.

The Advantages

● As the compression ratio increases, so does the torque, and the specific consumption is reduced correspondingly.
● The ignition characteristic map can be optimized with respect to power, consumption and exhaust.
● The ignition system automatically takes the engine knocking limit (as a function of the operating conditions) into consideration.
● Different fuel qualities, aging and environmental effects (pressure, temperature) are neutralized by the automatic selection of the correct ignition angle.
● Modern electronics makes it possible to carry out individual knocking detection, coupled with the appropriate knocking control, for each cylinder.

The closed-loop circuit for Knock-Control
The pressure signals from the pressure sensor are fed to the electronic control unit. If knocking vibrations occur, the control unit retards the ignition point until the engine operates again without knocking.

30

Ignition point
(Manipulated variable)

Knocking vibrations
(Controlled variable)

Engine
(Controlled system)

Ignition output stage (Final controlling element)

Knocking sensor

Electronic control unit

Control circuit

Evaluation circuit

31

Vibrogram showing a spark-ignition engine:
With knocking combustion

Vibrogram showing a spark-ignition engine:
Without knocking combustion

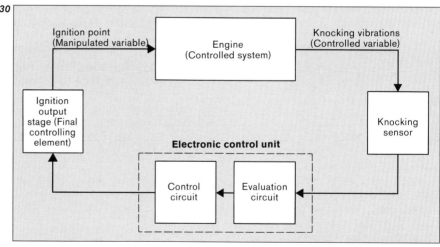

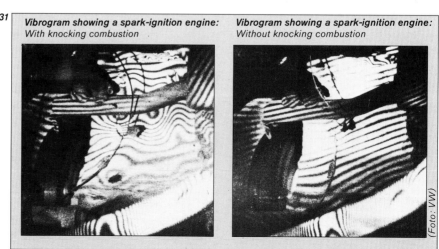

(Foto: VW)

Design

The knocking control system can either be combined as a separate unit with the transistorized ignition system or integrated in sophisticated systems such as the Motronic.

The possibility of combining electronic ignition and knocking control is particularly significant. A knocking sensor which is screw-mounted to the engine block is required for knocking control. The sensor consists of a piezo-ceramic disc[1] with a seismic mass[2] cast in plastic. In addition, a knocking-control circuit is required (as a separate closed-loop control unit or integrated into a sophisticated control unit).

Knocking control can be combined with the following functions:

1. Changing the air-fuel ratio by controlling the fuel pump, a deceleration fuel shutoff valve and an enrichment valve.
2. Exhaust gas recirculation.
3. Diagnosis, automatic recognition of trouble by the internal microprocessor and, if necessary, the output of information concerning the trouble by means of the tachometer (positioning the indicator at specific marks).

[1] Electric charges are created by pressure on piezo-ceramic material.
[2] A seismic mass acts as a conductor of deformations resulting from vibrations.

Operation

The engine knocking limits is not a fixed quantity but rather it depends on the various operating conditions. It is only important that the actual engine knocking limit be detected. The knocking sensor "listens" to the solid-borne vibrations from the engine block and converts these vibrations into electrical voltage signals.

The closed-loop control unit filters out and analyses the characteristic knocking signal. As shown in Figure 34, knocking detection and control are performed for each cylinder individually. The knocking signal is assigned to the appropriate cylinder. In this manner, it is possible to individually match the type of control used for each cylinder.

Once "knocking" has been detected, the control circuit immediately retards the ignition point of the next ignition for the appropriate cylinder (e.g. by 1.5° crankshaft). This process is repeated for each subsequent ignition if the sensor detects "knocking" again. This process is continued until the sensor no longer "hears" any knocking. Then, the control system advances the ignition point step by step to the value of the characteristic map which is closest to the engine knocking limit. In this manner, optimum engine efficiency

and fuel consumption are maintained. The signal of the control circuit triggers the ignition output stage. A safety circuit recognizes malfunctions and failures thus ensuring that the engine cannot be operated during a malfunction in an area where knocking is a danger. The ignition angle is retarded sufficiently and a warning indicator lights on the instrument panel.

The safety monitoring system has two circuits:

Circuit 1
The sensor, sensor lead and evaluation circuit are monitored.
Circuit 2
The microcomputer is monitored.

34) Knocking detection in the electronic control unit.
The characteristic knocking signal is filtered and analyzed from the signals supplied by the sensor. The microcomputer assigns the signals supplied by way of the AD converter to the appropriate cylinder and controls the ignition point of the next ignition for this cylinder.
▨ Control circuit ☐ Evaluation circuit

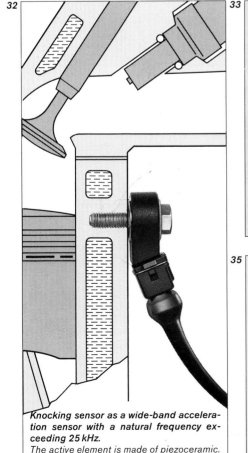

Knocking sensor as a wide-band acceleration sensor with a natural frequency exceeding 25 kHz.
The active element is made of piezoceramic. The sensor is surrounded with plastic to provide thermal isolation. The permissible operating temperature is 130°C.

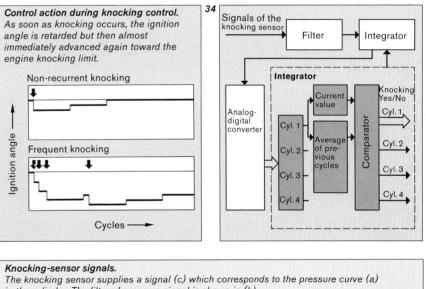

Control action during knocking control.
As soon as knocking occurs, the ignition angle is retarded but then almost immediately advanced again toward the engine knocking limit.

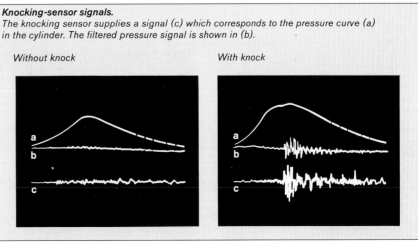

Knocking-sensor signals.
The knocking sensor supplies a signal (c) which corresponds to the pressure curve (a) in the cylinder. The filtered pressure signal is shown in (b).

Motronic

A combination of a breakerless semi-conductor-ignition system (characteristic-map ignition) and electronically controlled fuel injection (L-Jetronic) to jointly optimize both systems; designed using digital technology

The Problem

● Fuel injection and ignition should be controlled jointly and matched to one another, i.e. every operating condition must be sensed and taken into consideration during fuel metering and when determining the ignition point. This should result in the following improvements:
● Reduction of the fuel consumption.
● Increase in the specific engine power.
● Decrease in the high percentage of pollutants in the exhaust.

The Solution

An integrated system for electronically controlling fuel injection <u>and</u> ignition.

The Advantages

● Greater power.
● Higher fuel economy.
● Quieter running.
● Better reliability.
● Freedom from maintenance.
● Smooth engine operation.

The Ignition Subsystem

Calculation of the Ignition Angle

The microcomputer calculates the ignition angle for all operating conditions using the quantities of load, engine speed, temperature and throttle-valve position.

The ignition angle can be set to the optimum position for every operating point by means of the characteristic map stored digitally in the control unit. The engine speed is measured with high accuracy directly at the crankshaft by an induction-type speed sensor (22) (23). This helps reduce the offset from the engine knocking limit and to match better the ignition angle to the curve for maximum torque.

Dwell-angle Control

This is a function of the engine speed and battery voltage and is implemented using a dwell-angle characteristic map.

The Fuel-Injection Subsystem

Calculating the Duration of Injection

The bases for this are: The air-flow signal, the speed signal and correction factors. The basic injection duration is

calculated using these measured quantities for the engine load. The basic injection duration is then subjected to matching corrections.

The Fuel System

An electric fuel pump (2) sucks the fuel from the fuel tank (1). This pump generates the system pressure which is maintained at a constant pressure difference of 2.5 bar above the intake manifold pressure by means of an overflow pressure regulator (5). Since the pump supplies more fuel than is necessary, the excess fuel flows back into the fuel tank (1) by way of the vibration damper (28). The injection valves (10) are located in the individual intake manifolds such that they spray the fuel onto the heads of the injection valves. A thermo-time switch (18) controls the opening time of the cold-start valve.

Lambda Closed-loop Control

The Lambda closed-loop control (see Page 28) provides optimum matching of the air-fuel mixture. The Lambda sensor (17) is used to help set the mixture composition to the ideal value of $\lambda = 1$. This is done by measuring the exhaust-gas composition and appropriately changing the fuel supply. When $\lambda = 1$ pollutants are eliminated by a catalytic converter with a high degree of efficiency.

The System

36

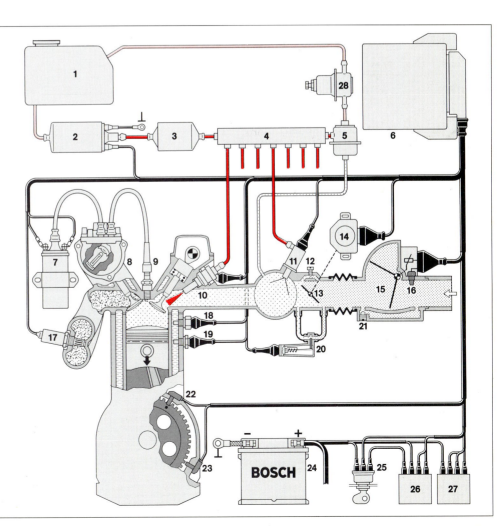

Overview of the Motronic System

1 Fuel tank
2 Fuel pump
3 Fuel filter
4 Fuel-distribution pipe
5 Pressure regulator
6 Control unit
7 Ignition coil
8 High-tension distributor
9 Spark plug
10 Injection valve
11 Cold-start valve
12 Idle-speed adjusting screw
13 Throttle valve
14 Throttle-valve switch
15 Air-flow sensor
16 Air-temperature sensor
17 Lambda sensor
18 Thermo-time switch
19 Engine-temperature sensor
20 Auxiliary-air valve
21 Idle-mixture adjusting screw
22 Reference-mark sensor
23 Engine-speed sensor
24 Battery
25 Ignition-speed sensor
26 Main relay
27 Pump relay
28 Vibration damper

The Control Unit

Design

The control unit consists of two printed boards which contain a microcomputer system. The computer system, in turn, consists of several large-scale-integration digital CMOS circuits.*)

The power components of the output stages for injection and ignition are mounted on the frame of the unit for better heat dissipation. A 35-pin connector is used to connect the control unit to the battery, the sensors and final-controlling elements.

Operation

Injection output stage

This stage controls the valve current which increases to the cut-in value but then drops to a holding current. An IC performs the control functions. The power section uses a Darlington transistor and a freewheeling transistor.

Ignition output stage

This stage essentially provides current amplification. The time for current to flow in the ignition coil is specified by the microcomputer as a function of the battery voltage and the engine speed. This stage also contains a control circuit for limiting the primary current of the ignition coil.

Pump control

The microcomputer controls switching of the fuel pump by means of an output stage and a relay.

Additional output stages

Depending upon the range of functions: Output stage for controlling the EGR valves,
Output stage for controlling a bypass valve for idle-speed control,
Output stage for controlling cylinder shutdown,
Output stage for controlling fuel-tank venting.

Output Signals

Injection signal

The microcomputer in the control unit calculates the injection duration as a function of the operating condition of the engine, and drives the output stage using the appropriate signal.

Ignition signal

The microcomputer in the control unit calculates an ignition angle using the input quantities of speed, load and various correction factors. In addition, the microcomputer determines the dwell angle. Both quantities are used to specify the duration of the ignition signal which is equal to the dwell period. This ignition signal controls the flow of current through the ignition coil by means of the ignition output stage.

*) CMOS: Complementary Metal-Oxide Semiconductor; MOS semiconductor technology using n-channel and p-channel transistors.

The Sensors
Input Signals

Engine speed (main control quantity)

Teeth of the induction-type pulse generator on the flywheel ring gear provide pulses for the control unit.

Crank angle

The induction-type reference-mark pulse generator on the flywheel ring gear registers pulses from the rotating reference mark and sends these pulses to the control unit.

Air quantity (main control quantity)

The angular position of the air-flow sensor flap with potentiometer provides a voltage signal for the control unit.

Signal Processing

Calculation of the duration of injection

The basic quantity of fuel injected is initially determined by the engine load.

Calculation of the ignition angle

This calculation uses the ignition characteristic map and the correction factors for starting, idle, overrun, part load and full load. These correction factors are a function of the engine temperature.

Calculation of the dwell angle

This angle is a function of the engine speed and the battery voltage.

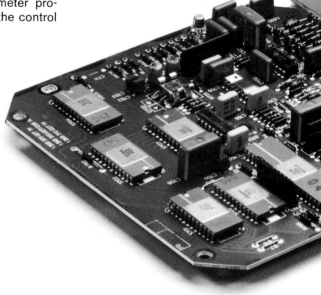

Block diagram of the control unit

37

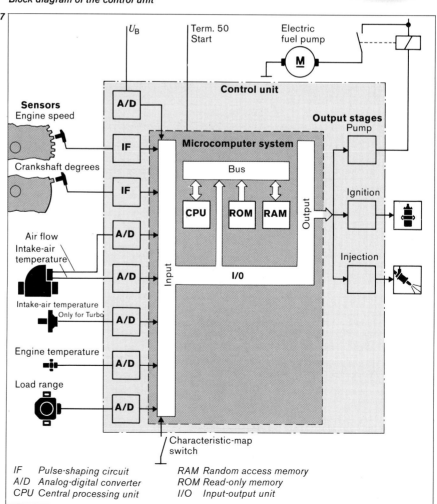

IF	Pulse-shaping circuit	RAM	Random access memory
A/D	Analog-digital converter	ROM	Read-only memory
CPU	Central processing unit	I/O	Input-output unit

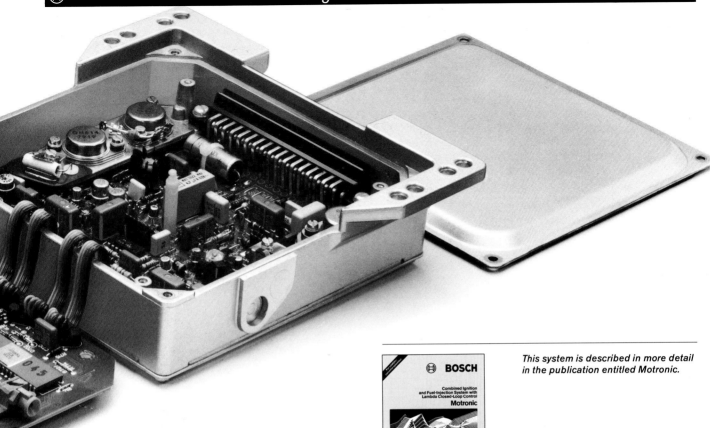

This system is described in more detail in the publication entitled Motronic.

BOSCH

Combined Ignition
and Fuel-Injection System with
Lambda Closed-Loop Control
Motronic

Technical Instruction

*Generation of the injection-duration pulse
(using a 6-cylinder engine as an example)*

38

Load sensing (air flow/speed) between 120° and 240° crankshaft	On / Off
Signal: Load-sensing completed	On / Off
Duration of injection is calculated using the correction factors	On / Off
The calculated duration of injection t_i is passed to the output stage	On / Off

0°   120°   240°   360°

*Generation of the dwell/ignition angle
(using a 6-cylinder engine as an example*

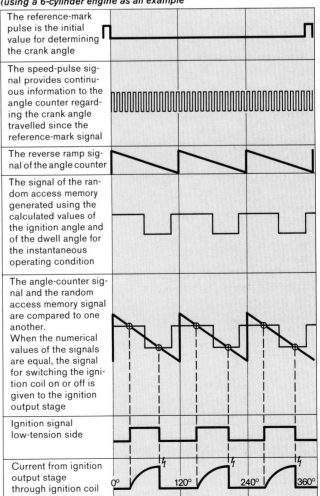

39

The reference-mark pulse is the initial value for determining the crank angle
The speed-pulse signal provides continuous information to the angle counter regarding the crank angle travelled since the reference-mark signal
The reverse ramp signal of the angle counter
The signal of the random access memory generated using the calculated values of the ignition angle and of the dwell angle for the instantaneous operating condition
The angle-counter signal and the random access memory signal are compared to one another. When the numerical values of the signals are equal, the signal for switching the ignition coil on or off is given to the ignition output stage
Ignition signal low-tension side
Current from ignition output stage through ignition coil

0°   120°   240°   360°

L-Jetronic

The Problem
- Reducing fuel consumption
- Increasing specific engine power

- Decreasing the high percentage of pollutants found in the exhaust.
- Improving driveability.

The Solution
Fuel injection as a function of the intake air flow and the engine speed.

The Advantages
- Increased fuel economy.
- Higher specific engine power.
- Improved driveability.
- Low-pollutant exhaust.
- Adaptable to every engine type.
- Maintenance-free.

Components
Fuel system
Tank, pump, filter, pressure regulator, distribution pipe, injection valves and cold-start valve.
Sensors
Air-flow sensor, air and engine temperature sensors, throttle-valve switch.
Control unit
Printed boards with electronic components, hybrid modules and IC.

Operation
Air-flow measurement
The air flow into the engine is the main control variable for determining the quantity of fuel injected. The intake air flows through the air-flow sensor (6) and deflects the moving sensor flap (6a) by a defined angle. This angle is converted into an electric voltage signal by a potentiometer and fed to the electronic control unit (7).
Electronic control of the quantity of fuel injected
The fuel is supplied by the electric fuel pump (2). Starting from the fuel tank (1), the fuel is routed via a fine-mesh filter (3) into a distribution pipe (4). The individual lines to the injection valves (9) branch off this pipe. A pressure regulator (5) at the distribution pipe maintains a constant system pressure. The control unit supplies the pulses for opening and closing the injection valves. The opening time of the valves specifies the quantity of fuel injected.
Optimum adaptation to various operating conditions
Cold starting: The cold-start valve (11) injects additional fuel into the intake manifold (10) during starting so that cold stating is ensured. The thermo-time switch (14) specifies the time during which the cold-starting valve remains open.

Warming-up: During the warm-up phase, the temperature sensor (8) provides for an increased fuel supply.

Full load: The throttle-valve switch (12a) corrects the air-fuel mixture to an optimum value for engines which are designed to run lean in the part-load range. This means that the mixture is enriched.

Idling: The auxiliary-air valve (13) supplies the engine with an increased quantity of air by bypassing the throttle valve. This increases the engine speed when the engine is cold. The control function is implemented using an electrically heated bimetallic element.

The System

Pressures and components of the L-Jetronic

- 🟥 System pressure
- 🟥 Suction line or return line
- Atmospheric pressure
- Intake manifold pressure

1 Fuel tank
2 Electric fuel pump
3 Fine-mesh filter
4 Distribution pipe
5 Pressure regulator
6 Air-flow sensor with flap (6a)
7 Control unit
8 Temperature sensor
9 Injection valve
10 Intake manifold
11 Cold-start valve
12 Throttle valve with switch (12a)
13 Auxiliary-air valve
14 Thermo-time switch
15 Ignition distributor
16 Relay set
17 Ignition-starting switch
18 Battery

BOSCH

The Control Unit

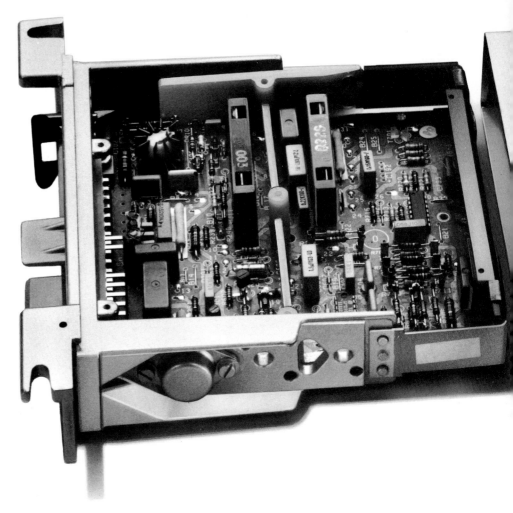

Design

Several assemblies are located on two printed-circuit boards. Three monolithic integrated circuits (ICs), together with the hybrid modules, form the heart of the entire circuit. Each IC has up to 120 integrated semiconductor components on its silicon chip. In addition, the printed boards hold several hundred discrete components such as transistors, diodes, capacitors, resistors and coils. The input and output leads are routed to the wiring harness by way of a multi-pin connector. The wiring harness provides the connections both to the sensors and switches as well as to the injection valves.

Operation

The control unit is the computer center for the electronic fuel-injection system. The control unit processes the input signals from the sensors and calculates the injection duration as a measure of the quantity of fuel to be injected.

Pulse-shaping circuit

This circuit converts the voltage pulses from the ignition system into rectangular pulses of the same frequency. The pulse-shaping circuit requires a pulse in order to become active but it switches itself off again automatically. This is how a monostable multivibrator works.

Frequency divider

This electronic functional unit is located on the same IC chip as the pulse-shaping circuit. It divides the frequency specified by the firing sequence so that two pulses are output for each complete rotation of the camshaft (Figure 44) regardless of the number of cylinders. The pulses are fed to the next stage (see below), the division-control multivibrator. Thus, the frequency divider halves the input frequency if the engine has four cylinders. In the case of a 6-cylinder engine, it divides the frequency by 3 and, for an 8-cylinder engine, by 4. The shape of the pulses is still rectangular. Since the frequency divider requires one pulse to switch on and another pulse to switch off, it is also known as a bistable multivibrator.

Division-control multivibrator

The speed data (n) and the air-flow signal (U_S) are the input signals to this circuit. The duration T_p (basic injection duration of its output pulse) of these pulses specifies the uncorrected quantity of fuel injected. Since the pulse duration is proportional to the intake air flow during each suction stroke, the air flow must be divided by the engine speed. This division is performed electronically by the charging/ discharging of a capacitor in the circuit.

Multiplier stage

The pulses of duration T_p control the multiplier stage. This stage collects additional data regarding various conditions of the engine such as warm-up, full load. From these conditions, the circuit calculates a correction factor which is multiplied by T_p. The charging/discharging of a capacitor is used in this case, too.

Output stage

Correction pulses for compensating the voltage for the injection valve are "attached" to the output pulses of the multiplier stage. These pulses are then used to control the output stage which is designed as a Darlington circuit. A current of roughly 1.5 A must be switched for each injection valve.

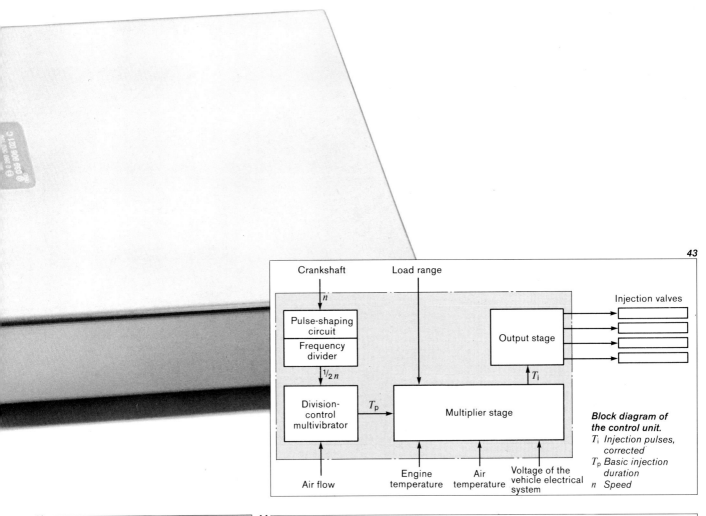

43

Crankshaft   Load range

Injection valves

n

Pulse-shaping circuit

Frequency divider

Output stage

$\frac{1}{2}n$

T_i

Division-control multivibrator

T_p

Multiplier stage

Air flow

Engine temperature   Air temperature   Voltage of the vehicle electrical system

Block diagram of the control unit.
T_i *Injection pulses, corrected*
T_p *Basic injection duration*
n *Speed*

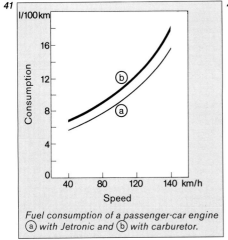

41

l/100km

16

12   (b)

8   (a)

4

0

40   80   120   140 km/h

Speed

Consumption

Fuel consumption of a passenger-car engine (a) with Jetronic and (b) with carburetor.

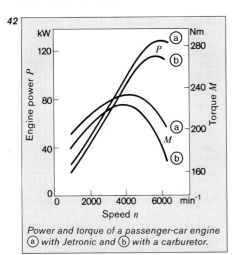

42

kW   Nm

120   P (a)   280
        (b)

            240

80

           (a)   200

40   M (b)

           160

0
0   2000   4000   6000 min⁻¹

Speed n

Engine power P Torque M

Power and torque of a passenger-car engine (a) with Jetronic and (b) with a carburetor.

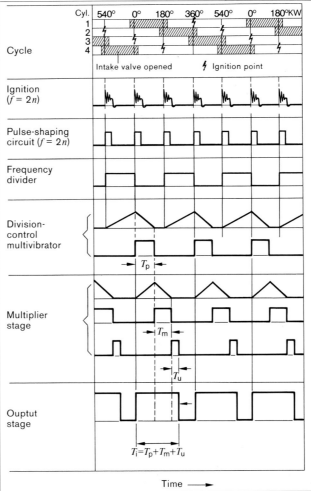

44

Cyl.   540°   0°   180°   360°   540°   0°   180°KW
1
2
Cycle   3
4

Intake valve opened   ⚡ Ignition point

Ignition ($f = 2n$)

Pulse-shaping circuit ($f = 2n$)

Frequency divider

Division-control multivibrator

$\leftarrow T_p \rightarrow$

Multiplier stage

$\leftarrow T_m \rightarrow$

T_u

Output stage

$T_i = T_p + T_m + T_u$

Time ⟶

Complete pulse diagram of the L-Jetronic for 4-cylinder engines
f *Ignition pulse frequency or firing sequence*
n *Engine speed*
T_p *Basic injection duration*
T_m *Increased pulse duration due to corrections*
T_u *Increased pulse duration due to voltage compensation*
T_i *Pulse control time The actual injection duration for each cycle differs from the pulse control time because the injection duration is changed both by a response delay and a drop-out delay.*

KW = crankshaft

LH-Jetronic
with Hot-wire Measurement of the Air Mass

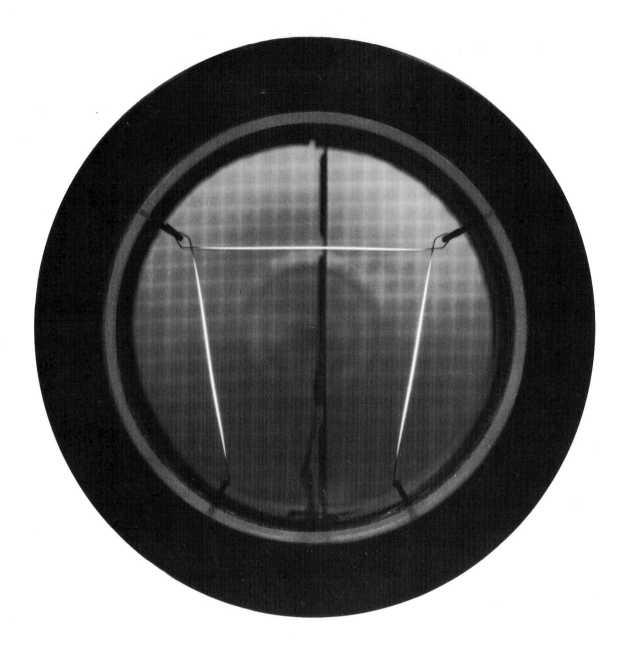

The Problem

When measuring the <u>quantity</u> of air sucked by the engine using an air-flow sensor flap, an error is introduced depending on the altitude. In addition, this type of measurement is prone to slight pulsation errors and the moving parts of the air-flow sensor are subject to wear.

The Solution

The air <u>mass</u> drawn in by the engine is measured directly by a hot-wire air-mass meter. This measurement is independent of changes in the density of the air.

The Advantages

● Precise measurement of the air mass.
● Fast response of the air-mass meter.
● Improved ability to adapt to engine operation.
Special advantages of the hot-wire air-mass meter.
● No measurement errors when driving at high altitudes.
● No pulsation errors.
● Fast response.
● No moving parts.
● Simple design.
● No errors in measurement due to a different intake-air temperature.

Design

A platinum wire, only 70 μm thick, is placed inside a measuring tube (Figure 45). The inner measuring tube consists of two plastic halves. The hot wire retainer ring, the precision resistor and the temperature sensor are mounted in these halves (Figure 45). The electronics containing the hybrid circuit, the power transistor and the idle potentiometer are located in the housing (Figure 45). The idle potentiometer is connected directly to the electronic control unit. The hybrid circuit contains some of the bridge resistors (bridge circuit, Figure 47) as well as the control and self-cleaning circuit.
The modular system was used for the design so that the components are combined into functional groups.

Operation

The hot-wire air-mass meter operates on the "constant temperature" principle. The hot wire is one arm of a bridge circuit (Figure 47). The diagonal voltage of this bridge is brought to zero by changing the heating current. As the air flow increases, the wire is cooled and the resistance drops. This changes the voltage relationships in the bridge circuit. The control circuit immediately corrects this situation by increasing the heating current. The increase in the current is such that the hot wire returns to original temperature again. This provides a defined relationship between the air flow and the heating current: The heating current is a measure of the air mass drawn into the engine.

The hot wire is regulated to constant temperature very quickly because of the low wire mass, and time constants of a few milliseconds are the result. This characteristic provides a major advantage: In the event of pulsations of the air (during full-load operation), the true air mass is measured so that pulsation errors such as those which occur in air-flow sensors are avoided. A measurement error occurs only if return flow is encountered. This situation arises at low engine speeds when the throttle valve is completely open. However, this measurement error can be offset using electronic means.

The hot-wire current is measured as the voltage drop across a precision resistor.

The resistance of the hot-wire and that of the precision resistor are designed so that the heating current ranges from 500 to 1200 mA, depending upon the air-flow rate. In the other arm of the bridge circuit, the current is only a fraction of the heating current because high-impedance resistors are used there. The same also applies to the temperature compensation resistor which has a resistance of around 500 ohms. This compensation resistor must maintain a constant resistance, be corrosion-resistant and exhibit rapid response. A platinum-film resistor was selected on the basis of these requirements. The compensation effect can be adjusted by means of the series resistor R1 (Figure 47). The temperature sensor is required to compensate for the intake-air temperature. Compensation must occur rapidly since the effect of temperature is very pronounced. Experiments have shown that a time constant of 3 seconds or less is necessary to ensure an accurate matching of the sensor output signal to the intake-air temperature. This is obtained because of the low mass of the sensor and of the connections.

Since the output signal can change if the surface of the hot wire becomes dirty, the hot wire is heated to an increased temperature for 1 second every time the engine is switched off. This action burns off any dirt on the hot wire. The "burn-off" command comes from the control unit of the LH-Jetronic.

The idle potentiometer which is also accommodated in the hot-wire air-mass meter is used for setting the idle mixture.

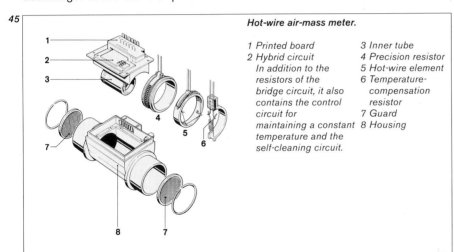

45

Hot-wire air-mass meter.

1 Printed board
2 Hybrid circuit
  In addition to the resistors of the bridge circuit, it also contains the control circuit for maintaining a constant temperature and the self-cleaning circuit.
3 Inner tube
4 Precision resistor
5 Hot-wire element
6 Temperature-compensation resistor
7 Guard
8 Housing

46) *The basic design of the LH-Jetronic is the same as that of the L-Jetronic. However, the hot-wire air-mass meter is used in place of the mechanical air-flow sensor. It supplies the air-flow signal.*
The idle actuator is a controlled bypass of the throttle valve. It supplies a specific quantity of air to the engine during idling. This quantity of air is specified by a microcomputer in the control unit in accordance with the input signals.

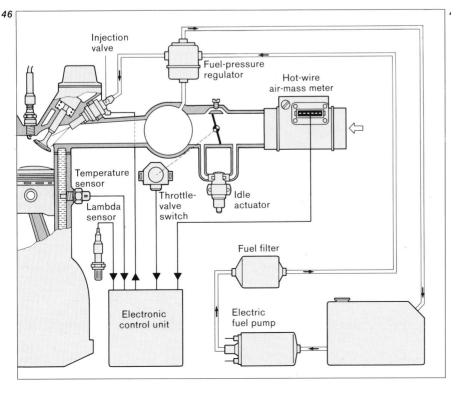

46

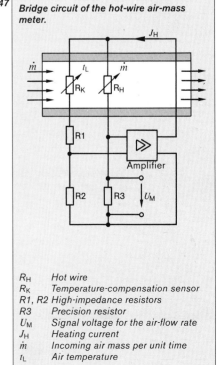

47

Bridge circuit of the hot-wire air-mass meter.

R_H Hot wire
R_K Temperature-compensation sensor
R1, R2 High-impedance resistors
R3   Precision resistor
U_M Signal voltage for the air-flow rate
J_H Heating current
$\dot{m}$ Incoming air mass per unit time
t_L Air temperature

KE-Jetronic

The Problem

Matching the mixture to changing loads and operating conditions is to be improved compared to the K-Jetronic.

The K-Jetronic cannot process the engine temperature. Rapid load-change commands from the throttle valve are detected only after a delay, and a Lambda closed-loop control circuit requires great additional expenditure.

The Solution

An additional electronic control system makes expanded sensing of operating data possible.

This makes warm-up control with post-start enrichment possible.

In particular, acceleration performance is improved. It is feasible to expand the system with other "intelligent" functions in the area of emission control.

The Advantages

● Improved driveability during warm-up because of the short time required to respond to movements of the accelerator pedal.
● Improved fuel economy.
● Additional reduction in fuel consumption provided by the overrun fuel cut-off.
● Reduction of pollutants in the exhaust.

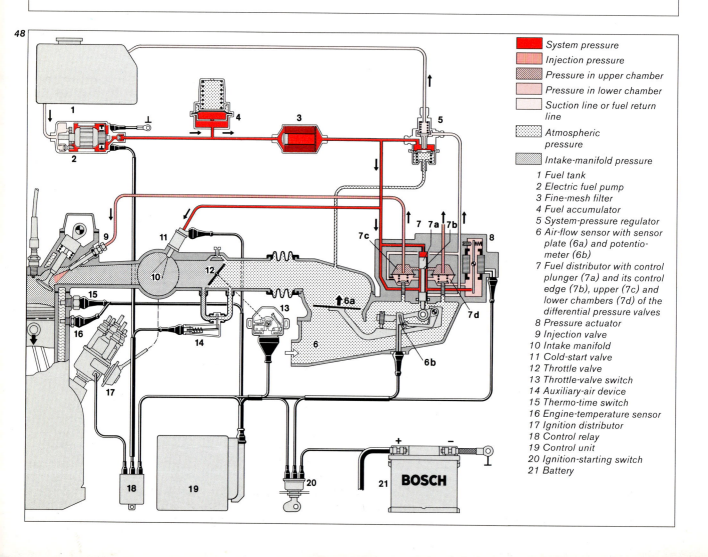

48

System pressure
Injection pressure
Pressure in upper chamber
Pressure in lower chamber
Suction line or fuel return line
Atmospheric pressure
Intake-manifold pressure

 1 Fuel tank
 2 Electric fuel pump
 3 Fine-mesh filter
 4 Fuel accumulator
 5 System-pressure regulator
 6 Air-flow sensor with sensor plate (6a) and potentiometer (6b)
 7 Fuel distributor with control plunger (7a) and its control edge (7b), upper (7c) and lower chambers (7d) of the differential pressure valves
 8 Pressure actuator
 9 Injection valve
10 Intake manifold
11 Cold-start valve
12 Throttle valve
13 Throttle-valve switch
14 Auxiliary-air device
15 Thermo-time switch
16 Engine-temperature sensor
17 Ignition distributor
18 Control relay
19 Control unit
20 Ignition-starting switch
21 Battery

Components

Air-flow sensor with sensor plate and potentiometer.
Fuel distributor with electro-hydraulic pressure actuator.
Electric start valve. Throttle-valve switch and auxiliary-air device.
Thermo-time switch and engine-temperature sensor.
Electric fuel pump, fuel filter, fuel accumulator, pressure regulator and injection valves.
Electronic control unit.

Operation

The KE-Jetronic is based on the mechanical K-Jetronic and was provided with an electronic system for the purpose of increased flexibility and to implement "intelligent" functions (the "E" in KE stands for Electronic).
The basic system of the KE-Jetronic operates mechanically.
A diaphragm pressure regulator provides the system and control pressures.
The control system acts centrally by means of an electrohydraulic final controlling element. This element changes the fuel metering in the fuel distributor. The response times to changes in the position of the accelerator pedal are extremely short.
Because the basic system is mechanical, the KE-Jetronic can continue to run under emergency ("limp-home") conditions.
In addition to the advantages provided by improved functions, the units of the KE-Jetronic are small in size and low in weight.
The hydraulic subsystem of the KE-

Jetronic is the same for both Lambda-controlled systems and those which are not controlled. The electronic control unit for the KE-Jetronic with closed-loop Lambda control is designed for the required additional function.

Correction Functions

When starting the engine, the air-fuel mixture – depending on the starting temperature – is enriched up to a factor of 2.5 ($\lambda \approx 0.4$) thus rapidly completing any required filling operations in the system.
The warm-up enrichment is a function of the coolant temperature. This temperature is measured by a NTC temperature-sensor resistor.
The post-start enrichment is also dependent on temperature: Starting from an initial value, the enrichment is regulated in accordance with a linear function of time.
The maximum value of the acceleration enrichment is a function of temperature. When the position of the accelerator pedal is changed rapidly, a maximum factor of 1.7 times the stoichiometric air-fuel ratio is used for enrichment purposes while for slow changes in the accelerator pedal position, the factor is reduced to 1.1 times. The rate at which the accelerator pedal is changed is derived from the motion of the air-flow sensor plate which exhibits only a slight delay with respect to the motion of the throttle valve. This signal, which corresponds to the change in engine power over time, is sensed by the potentiometer in the air-flow sensor and processed in the electronic control unit. The potentiometer

characteristic is non-linear so that the acceleration signal is a maximum when moving from the idle position and decreases with increasing engine power. Since "natural" full-load enrichment in 4-cylinder engines is highly dependent on engine speed (due to air pulsation), enrichment is matched in a complementary fashion to the natural behavior by the KE-Jetronic. This results in virtually constant full-load CO over the entire speed range thus considerably facilitating matching action.

Additional Functions

Overrun fuel cut-off takes effect during overrun, provides smooth transition, and responds depending on coolant temperature. Engine-speed data are taken from the ignition system. If the engine is warm, the switching thresholds are set as low as possible so that a lot of fuel can be saved. If the coolant temperature is low, threshold values increase so that the cold engine does not stall when the clutch pedal is depressed (manual transmission) or the accelerator is released (automatic). The following functions have also been provided for and a few have already been implemented:
In systems with Lambda closed-loop control, rate of control is matched to engine power. This rate increases from the minimum value during idle to three times minimum value once engine power exceeds ten times idle power. The same control current used for overrun fuel cut-off can also be used to stop fuel injection thus providing an engine-speed limit. Additional functions can be input by means of signals and are processed there to provide an appropriate control current for the final controlling element.
Altitude correction takes into consideration the changing air density with increasing altitude. The air-fuel mixture becomes leaner as a function of atmospheric pressure. Pressure is measured using an aneroid box. The signal is provided by a potentiometer.

The Control Unit

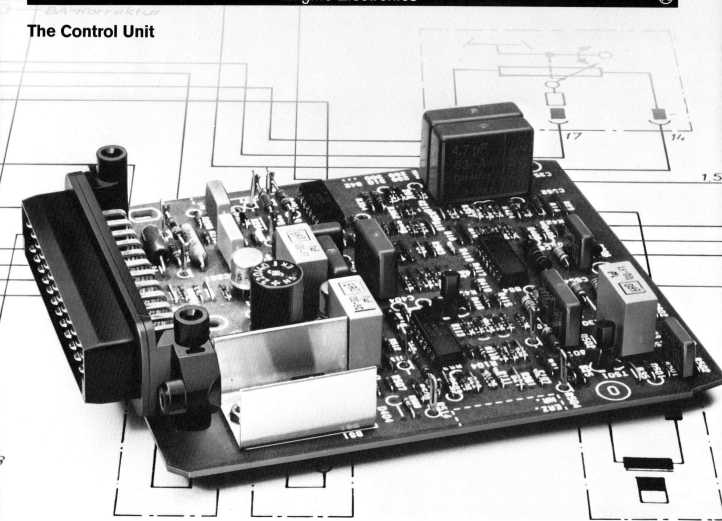

Design

The electronic components of the KE-Jetronic control unit are linear integrated circuits (operational amplifiers, comparators, and a voltage regulator), transistors, diodes, resistors and capacitors. These components are mounted on one printed board.

The control unit is designed as a plug-in device and is accommodated in a housing provided with a pressure equalization element.

Operation

The control unit must supply a control current as the output quantity on the basis of the input quantities coming from various sensors. This control current is used to drive the electro-hydraulic final controlling element (which is continuously adjustable). The control action affects the differential pressure in the fuel distributor[1].

Voltage regulation

The control unit requires a constant voltage. This means it must be maintained constant regardless of the voltage provided by the vehicle electrical system. This voltage is used to provide the current for the final controlling element, this current being a function of the engine status quantities. The voltage for the control unit is regulated in an integrated circuit.

Input filters

These circuits filter any noise out of the input signals.

Voltage increase for starting

When the starter motor is operated, the control unit provides a "start current" which is sent to the actuator as the maximum current for approx. 1.5 s.

Post-start enrichment

Following the starting phase, the fuel is enriched depending on the engine temperature. This fuel enrichment is maintained at its maximum value by the control unit for roughly 4.5 s and is then decreased over approximately 20 s (this applies to starts at 20 °C).

Warm-up enrichment

An NTC sensor provides a voltage, as a function of temperature, which is then converted into an appropriate current for the final controlling element.

Acceleration enrichment

During positive load changes, the engine, primarily when cold, requires an increased fuel quantity for a brief period. The magnitude of this extra fuel quantity is dependent on the engine temperature and the rate of opening, as well as on the initial position of the throttle valve (non-linear potentiometer characteristic).

The control unit processes the rate of change of the voltage from the potentiometer connected to the air-flow sensor plate and reduces the enrichment over 1 ... 3 s.

Deceleration fuel shutoff

During the overrun phase (the cut-off speed has been exceeded), and with the throttle-valve switch closed, the control unit closes the differential-pressure valves in the fuel distributor by means of the pressure actuator. This stops fuel injection. As the engine speed drops, fuel injection commences again when the reinstatement speed is reached.

Full-load correction

This is controlled by the full-load contact on the throttle valve and by the engine speed. Full-load correction is used to compensate for the situation of the air-fuel mixture becoming leaner or richer due to air pulsations in the intake-air flow.

Summing circuit

This circuit combines the correction signals and operates as a controller. The correction signals are summed in an operational amplifier circuit and then fed to the current controller.

Output stage

This circuit converts the controller output signal or the overrun fuel cut-off signal into a drive current for the actuator.

[1] See the Bosch Technical Instruction for the K-Jetronic on this topic

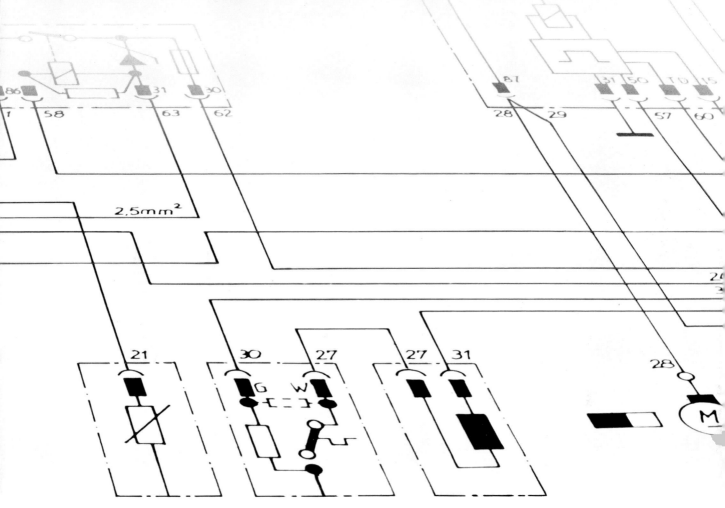

It is possible for the output stage to act as either a source or a sink of current for the final controlling element.

With a transistor operating in the active range, the output stage acts as a source for the current to the actuator and can provide any current desired. During overrun (overrun fuel cut-off), the output stage acts as a current sink. This current affects the differential pressure such that the fuel supply is shut off.

A detailed description of the K-Jetronic basic system can be found in the Technical Instruction Publication entitled K-Jetronic.

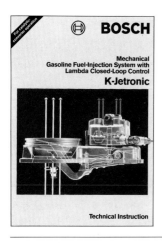

49

Block diagram of the KE-Jetronic control unit.
The correction signals from the various blocks are combined in the summing circuit, amplified in the output stage and fed to the electrohydraulic final controlling element.

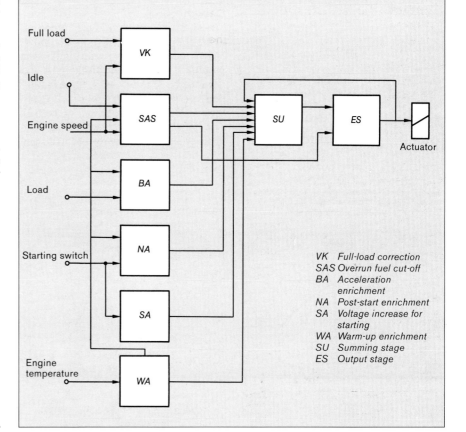

VK Full-load correction
SAS Overrun fuel cut-off
BA Acceleration enrichment
NA Post-start enrichment
SA Voltage increase for starting
WA Warm-up enrichment
SU Summing stage
ES Output stage

Lambda Closed-Loop Control

The Problem

The level of pollutants in exhaust gas must be reduced. While open-loop control systems for ignition and fuel management can improve exhaust emissions, a reduction in the pollutant emission exceeding this is only possible by using catalytic converters. These converters can only operate efficiently if unleaded gasoline is used and if combustion is as complete as possible.

The Solution

The Lambda closed-loop control system provides <u>closed-loop control</u> of the injected quantity of fuel such that the air-fuel mixture enables virtually complete combustion to take place.

The Advantages

By providing closed-loop control of the mixture, it becomes possible to use catalytic converters for the exhaust and thus to achieve the maximum improvement in exhaust gas possible at the present time. Fuel comsumption is also reduced.

Mixture Composition

The term "excess-air factor" λ (Lambda) was selected to designate the air-fuel mixture.

$$\lambda = \frac{\text{Supplied quantity of air}}{\text{Theoretical air requirement}}$$

$\lambda = 1$: The supplied quantity of air is equal to the quantity of air theoretically required.
$\lambda < 1$: Insufficient air or a "rich" mixture.
$\lambda > 1$: Excess air or a "lean" mixture.
The power, fuel consumption and exhaust-gas composition of a spark-ignition engine are highly dependent on the composition of the air-fuel mixture. Complete combustion occurs when using gasoline at an air-fuel ratio of roughly 14:1 (14 kg air: 1 kg gasoline) ($\lambda = 1.0$).
Figures 50 and 51 show the output of pollutants and fuel consumption as a function of the excess-air ratio. The best values for CO and CH and for fuel consumption are obtained in the area of $\lambda = 1$.

The Probe

Design

The ceramic sensor body is contained in a housing which protects it against mechanical effects and facilitates mounting. The outer section of the ceramic body is placed in the exhaust-gas stream, the inner section is in contact with the ambient air (Figure 52). The ceramic body is essentially made of Zirconium dioxide. Its surfaces are provided with electrodes made of a very thin gas-permeable platinum lay-er. In addition, a porous ceramic coating has been applied to the side exposed to the exhaust gas. This coating prevents contamination of the electrode surface by combustion residue in the exhaust-gas stream and ensures uniform probe properties.

Operation

The Lambda sensor is used to measure the oxygen content in the exhaust gas. Since oxygen components are still present in the exhaust gas even with a rich mixture, the sensor signal is a measure of the mixture composition. The principle of sensor operation is based on the fact that, starting at temperatures of roughly 300 °C, the ceramic material used conducts oxygen ions. If the percentage of oxygen present on both sides of the sensor is different, a step-like voltage curve in the area of $\lambda = 1$ is generated on the basis of the special properties of the material used (Figure 53).

The Closed-loop Control

In the Jetronic and the Motronic, the closed-loop control is performed by a special module of the control unit.
It has the following functional groups: sensor monitor, controller with signal amplifier and integrator.

Probe Monitor

This circuit is to monitor whether the Lambda sensor is operating correctly in conjunction with the closed-loop control circuit. The sensor is not operating correctly if, for example, the ceramic sensor body is at a temperature of less than 300 °C because, in this case, the electrical resistance of the ceramic sensor body is too high for

---

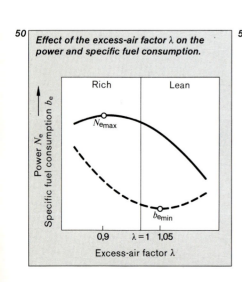

50   *Effect of the excess-air factor λ on the power and specific fuel consumption.*

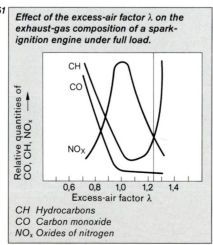

51   *Effect of the excess-air factor λ on the exhaust-gas composition of a spark-ignition engine under full load.*

CH Hydrocarbons
CO Carbon monoxide
NO_x Oxides of nitrogen

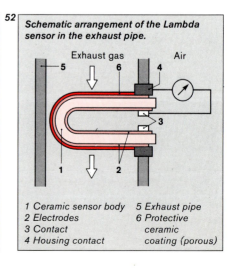

52   *Schematic arrangement of the Lambda sensor in the exhaust pipe.*

1 Ceramic sensor body
2 Electrodes
3 Contact
4 Housing contact
5 Exhaust pipe
6 Protective ceramic coating (porous)

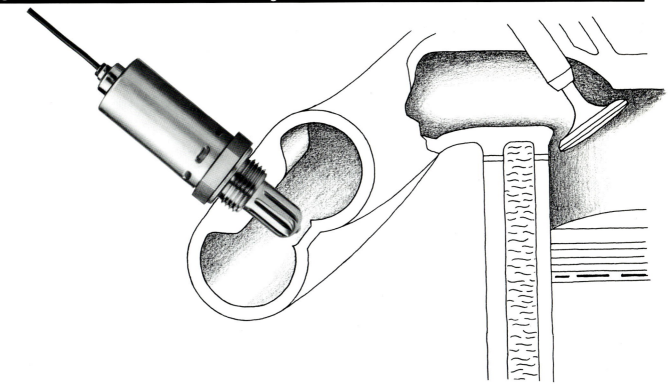

correct evaluation of the sensor voltage U_λ. Under such conditions, the sensor monitor opens the closed loop. In this case, the fuel metering equipment receives a constant signal from the controller. This signal results in a mixture composition for average driving. This controller instruction takes precedence over the start and warm-up enrichment of the fuel metering equipment.

The sensor monitor takes the same action if the lead between the sensor and the controller breaks.

Controller

In the case of a "rich" mixture, $U_\lambda > U_r$ (where U_r is a specified reference voltage). In this case, the signal amplifier outputs no voltage to the other functional groups in the controller. During this time (T_0) when no pulses are output, less fuel is supplied to the engine and the air-fuel mixture becomes leaner and leaner. If the mixture now moves into the range of $\lambda > 1$ ("lean"), the sensor voltage U_λ drops

below the reference voltage U_r and the signal amplifier is switched on. This means that the signal amplifier outputs amplified voltage pulses having a period of T_1.

The voltage pulses are fed from the signal amplifier to the integrator. In this circuit, the pulses are changed to provide an indication of the control tendency. For, as soon as U_λ exceeds or drops below the voltage U_r, the air-fuel mixture begins to change but not abruptly. Instead, the control system begins to act in the appropriate direction at a rate programmed by the integrator so as not to have a negative effect on driving.

As a result of this, the control system is constantly changing the composition of the air-fuel mixture within a narrow tolerance band around $\lambda = 1$ from the direction "rich" to "lean" and back (Figure 54).

The Closed-loop Control Circuit

The closed-loop control circuit consists of the engine (the controlled system), the Lambda sensor (the measuring element), the controller in the control unit and the injection valves (the final controlling element) (Figure 55).

Operation

The fuel supply to the engine is regulated by the fuel management system in accordance with the mixture-composition data provided by the Lambda sensor. This control is designed so that the air-fuel ratio deviates from $\lambda = 1$ by only a small amount. Control is effected by means of the quantity of fuel injected at the injection valve (final controlling element). The control system automatically takes special operating conditions such as starting, acceleration and full load into consideration.

---

53 *Voltage curve for the Lambda sensor at an operating temperature of 600°C.*

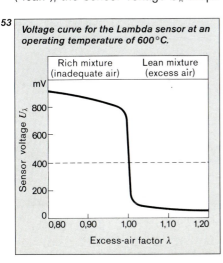

54 *Timing diagram of the Lambda closed-loop control.*

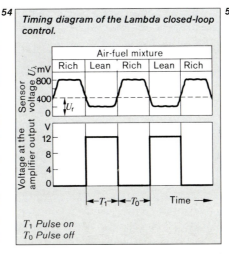

T_1 Pulse on
T_0 Pulse off

55 *Functional diagram taking as an example the L-Jetronic equipped with Lambda closed-loop control.*

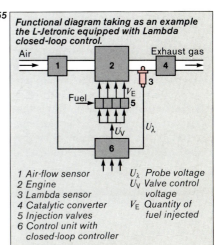

1 Air-flow sensor
2 Engine
3 Lambda sensor
4 Catalytic converter
5 Injection valves
6 Control unit with closed-loop controller

U_λ Probe voltage
U_V Valve control voltage
V_E Quantity of fuel injected

Idle-Speed Control

Idle-speed rotary actuator

56)　　　　　　　　　　　*Throttle-valve switch*

The Problem

Changes in engine temperature and thus changes in the engine friction, together with engine aging – all these factors cause the idle speed to vary if the cross section of the bypass is constant.

The Solution

The idle speed is controlled by measuring the idle speed and by an electro-mechanical change of the bypass cross-section.

The Advantages

● Constant, low-emission and economical idling is ensured under the most varied conditions.
● Substantially lower idle speeds can be used than for the same engines without closed-loop idle control. This saves fuel.
● Abrupt drops in speed which occur when connecting loads such as power-assisted steering, putting the automatic transmission in drive, or switching on the air conditioner are eliminated.
● It is possible to increase or decrease the idle speed when the air conditioner is switched on or the automatic transmission is in "drive".
● The function performed by the auxiliary-air device of Jetronic systems can be performed by the idle-speed (closed-loop) control.

The System

The idle-speed control can be used only in conjunction with a gasoline injection system. It has the following components:
Idle-speed actuator
Throttle-valve switch
Idle-speed switch
Temperature sensor
Closed-loop control unit (idle-speed controller).

Idle-speed Actuator

The actuator is installed in the bypass tubing of the throttle valve (in place of an auxiliary-air device). The opening cross-section of the actuator determines the idle speed of the engine.

Temperature Sensor

When cold, the engine reacts much better if the idle speed is higher. The signal required for this is supplied by a temperature sensor installed in the coolant circuit. As the coolant temperature increases, the idle speed is reduced to the "warm value" either gradually or abruptly.

Throttle-valve Switch

If the engine speed is increased by opening the throttle valve, the control unit attempts to regulate the system back to the set speed. This means that the actuator moves in the "close" direction until it is stopped electrically at the end point. However, this situation must be prevented, otherwise if the throttle valve is closed and an additional load (automatic transmission, air conditioner, power steering) is connected at the same time, the speed would drop drastically for a short time.

For this reason, a signal from the throttle-valve switch causes the minimum opening cross-section on the actuator to be increased when the throttle valve is open. Due to this, the control action starts with the actuator open thus avoiding a drop in speed.

Idle-speed Switch

In order to be able to provide the required cooling power of the air conditioner, it may be necessary to increase the idle speed. It is also often necessary to reduce the idle speed when drive is engaged in vehicles with an automatic transmission.
The increase or reduction of the idle speed is implemented using the idle-speed switch.

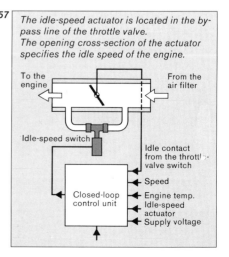

57) *The idle-speed actuator is located in the bypass line of the throttle valve.*
The opening cross-section of the actuator specifies the idle speed of the engine.

The Closed-loop Control Unit

The closed-loop control unit contains the following **function blocks:**
Frequency (speed) – voltage converter, a block for the set speed, a block for dropping speed, an insensitive range block, a proportional-integral controller, a block for limiting the control range, an oscillator, a pulse-width modulator and an output stage.

Operation
The set speed is regulated by the closed-loop control unit. The required data regarding the actual speed is fed to the control unit from the ignition system. The pulses corresponding to the speed are converted to a voltage signal in the control unit. This signal is compared to a voltage corresponding to the set speed. From the voltage difference, the control unit generates the actuator signal which is fed to the idle-speed actuator as a pulse train. The oscillator and the pulse-width modulator are used to generate this pulse train.
The insensitive-range block reduces the control gain near the set speed. The controller operates as a so-called proportional-integral controller (PI-controller). This means that the P section of the controller converts the control deviation into an actuating signal which is proportional to the deviation both with respect to time and magnitude. The I-section processes abrupt changes in the input signals, and spreads them with respect to time in the output signals.
The actuating signal is the sum of both signal processing characteristics.
The control unit is designed such that, according to the type of engine, the minimum opening cross-section of the actuator is electrically limited so that it generally cannot close completely.
Additional inputs to the control unit such as those from the temperature sensor and the throttle-valve switch

58) *The electronic closed-loop control unit (idle controller) for the idle-speed control.*

are provided so that unwanted control action is avoided even under special conditions, for example, at low temperatures, during changes in the speed or when "stepping on the gas".

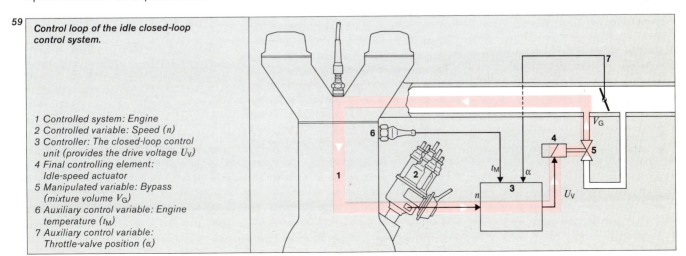

59

Control loop of the idle closed-loop control system.

1 Controlled system: Engine
2 Controlled variable: Speed (n)
3 Controller: The closed-loop control unit (provides the drive voltage U_V)
4 Final controlling element: Idle-speed actuator
5 Manipulated variable: Bypass (mixture volume V_G)
6 Auxiliary control variable: Engine temperature (t_M)
7 Auxiliary control variable: Throttle-valve position (α)

ECOTRONIC

**Bosch-Pierburg Carburetor
with Electronically
Controlled Functions**

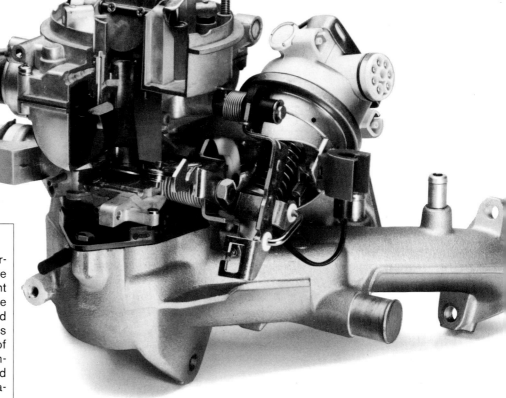

The Problem

The design of the mechanical carburetor has become more and more complicated with the development of high-speed engines, the increased demands placed on cold starting and on warm-up, as well as the more stringent requirements of emission regulations. These complex tasks cannot be performed using conventional, purely mechanical means.

The Solution

Advanced micro electronics makes it possible to equip simplified carburetors with electrical control elements. These adapt the carburetor functions, with a high degree of accuracy, to obtain the best-possible figures for fuel consumption and exhaust emissions for the particular operating conditions.

The Advantages

● Sure starting, both cold and warm.
● Smooth running of the engine.
● Stable idling at low engine speeds.
● Smooth, progressive acceleration.
● Reliable engine transition following overrunning and braking.
● Integrated overrun fuel cut-off.
● Integrated idle-speed control.
● No dieseling after stopping the engine.
● Reliable "limp-home" features.
● A central electronic control unit can be used.
● Already existing sensors can be used.
● Expansion potential by the addition of other input signals.

The System

Design

This electronic mixture-formation system for spark-ignition engines consists of a carburetor simplified to its basic systems and equipped with final controlling elements, an electronic control unit and sensors.

Carburetor with Final Controlling Elements

The carburetor housing contains the float system, the throttle valve and the throttling orifice (the choke valve), the idle system and the main nozzle system.

The throttle valve is opened to a greater or lesser degree by means of an electro-pneumatic final controlling element.

If the driver does not actuate the accelerator, the engine cylinder-fill is matched to the driving requirements. Another electro-magnetic final controlling element is used to close the choke valve and enriches the intake mixture with fuel to the degree desired.

Electronic Control Unit

The final controlling elements are regulated by means of an electronic control unit which consists of three function blocks: The input section, the output section and a computational block.

The Sensors

The electronic control unit is provided with signals from the sensors for the engine speed, engine temperature and/or intake-manifold wall temperature, the throttle-valve position, the position of the throttle-valve actuator (return signal), the idle-speed switch and any other sensors such as a Lambda sensor.

Operation

The system performs the following functions:
– Control of the mixture enrichment during starting, warming-up and acceleration
– Regulation of the idle speed
– Overrun fuel cut-off.

Control of the Mixture Enrichment during Starting, Warming-up and Acceleration

Depending upon the measured operating parameters, the electronic control unit selects the setting of the choke valve for mixture enrichment. It also selects the positioning of the main throttle for controlling cylinder-fill. Because of this, short starting times are achieved and the engine reliably runs up to speed at any ambient temperature. It also provides lean matching with good performance during warm-up.

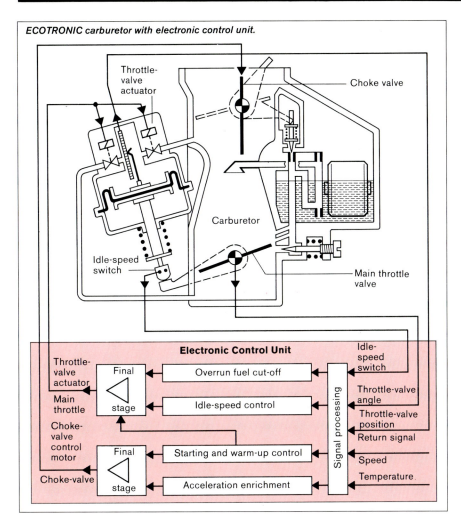

ECOTRONIC carburetor with electronic control unit.

speed threshold of 1100 ... 1400/min. It can be even higher in city driving or highway driving.

The overrun fuel cut-off does not have any negative effects on exhaust emissions of carbon monoxide, unburned hydrocarbons and oxides of nitrogen or on the performance of the vehicle.

Summary
The ECOTRONIC system makes it possible to save fuel, especially in city driving. This is done with precise control of the air-fuel ratio and cylinder-fill under varying operating conditions. It is easier to optimize the engine in terms of fuel consumption, exhaust emissions and performance. It also allows the system to be expanded with additional electronic functions for purposes of engine control.

The Control Unit

The electronic control unit is designed using digital technology. It can be subdivided into an input section, a processing section and an output section. It contains a 8-bit microprocessor[1].

The input section of the unit generates the supply voltages for the sensors (throttle-valve angle, temperature, idle setting), and digitizes the signals coming from the sensors. The processing section calculates the output values using the input variables together with pre-programmed operational curves. In order to be able to handle all cases, these operational curves are broken down into data points and linear interpolation is used to represent the curves between these data points. The output values of the processing section control the actuators of the choke valve and the main throttle valve after being converted in the power output stage of the output section.

Since the operating conditions such as the ambient temperature and the voltage supplied by the vehicle electrical system vary greatly, the design of the control unit incorporates features which ensure its functional reliability under all conditions occuring in practice.

The high rate of actuation of the choke-valve control motor makes it possible to use the choke valve for acceleration enrichment.

The throttle-valve-angle sensor detects acceleration. The microcomputer briefly closes the choke valve. In order to obtain the desired degree of enrichment, the required control of the choke-valve actuator is calculated as a function of the operating parameters of engine temperature, speed, throttle valve position, and the rate of opening of the throttle valve both during and shortly after the acceleration phase.

Since the simplified carburetor is tuned to the desired mixture ratio in stationary operation with the engine at working temperature, the electronic carburetor can operate well under emergency conditions over the entire range of the characteristic map even if the electronic control system fails.

Regulation of the Idle Speed
The measured engine speed is constantly compared to the desired speed and a control device calculates the required angle of the throttle valve using the deviation between these two speeds. The controller operates in idle at all engine temperatures and ensures a constant idle speed regardless of the frictional torque of the engine and of the torque requirement

when the engine is loaded by any other equipment in the vehicle.

In general, the control system allows the idle speed to drop since extra speed to accommodate the aforementioned effects is not necessary.

The control action to change the cylinder-fill takes place by positioning the throttle valve. The throttle-valve actuator contains two solenoid valves for connecting the control pressures (atmospheric pressure and intake-manifold pressure) and a potentiometer for sensing the position of the throttle-valve actuator.

Overrun fuel cut-off
During overrun, the engine does not output any power. Therefore, it does not require any fuel during this phase of operation. Fuel is shut off at speeds above the "overrun speed threshold" by closing the main throttle valve during overrun until it is below the idle position. In this manner, the idle-mixture outlet above the closed throttle valve extends into the area of atmospheric pressure so that the fuel supply from the idle system is stopped completely. This method is also used to prevent so-called "dieseling" when the engine is shut off.

The improved fuel economy achieved by this system is roughly 5% with a normal idle-speed setting and a cut-in

[1] 8-bit: 8-position binary number using the numerals 0 and 1, e.g. 10 101 110

Electronically Controlled Mixture Pre-heating for Diesel Engines

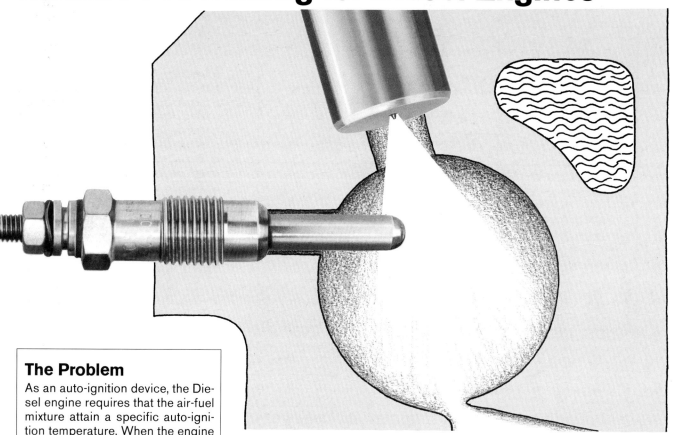

The Problem

As an auto-ignition device, the Diesel engine requires that the air-fuel mixture attain a specific auto-ignition temperature. When the engine is cold, this temperature cannot be achieved by compression alone. The injected fuel must be vaporized and ignited. This can be done either by heating the air or by the use of glow plugs. Long pre-heating times put a large drain on the battery. Plugs with short pre-heating times require precise timing control.

The Solution

Advanced systems for mixture pre-heating in Diesel engines are equipped with electronic control units. In place of manual actuation of the switches, these units have an electronic control system for the pre-heating time of the glow plug or flame plug.

The Advantages

● Short pre-heating times.
● Post-heating times chosen as a function of the operating conditions to provide complete, low-emission fuel combustion.
● Saving of the battery by the timing safety switch.
● Improved ease of operation, with starting just like a spark-ignition engine.

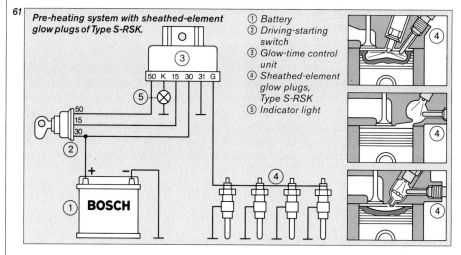

61 Pre-heating system with sheathed-element glow plugs of Type S-RSK.

① Battery
② Driving-starting switch
③ Glow-time control unit
④ Sheathed-element glow plugs, Type S-RSK
⑤ Indicator light

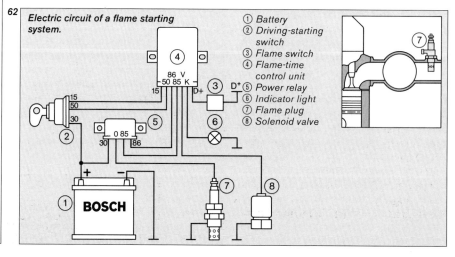

62 Electric circuit of a flame starting system.

① Battery
② Driving-starting switch
③ Flame switch
④ Flame-time control unit
⑤ Power relay
⑥ Indicator light
⑦ Flame plug
⑧ Solenoid valve

The System

Sheated-element Glow Plug in the Combustion Chamber

This system consists of the following components: Sheathed-element glow plugs, driving-starting switch, glow-time control unit, indicator light and battery (Figure 61).

The sheathed-element glow plug is screw-mounted in a specific location in the cylinder head depending upon the engine type. The plug consists of a shell into which a heating element is pressed so as to be gastight. Inside the heating element there is a helical heating filament which is electrically isolated in a powder packing. This filament consists of a typical heating resistor and a PTC resistor.

The helical heating filament in the glow plug is heated by means of an electric current. It transfers its heat to the powder packing which, in turn, transfers its heat uniformly to the glow tube. Depending upon the material used and the shape, glow plugs reach the temperature necessary to ignite the air-fuel mixture in approximately 4 to 10 seconds (with an ambient temperature of 0 °C). These temperatures are roughly 900 °C.

The fuel jet injected into the combustion chamber does not directly strike the glow plug because fuel striking the glowing plug surface would destroy this surface. The hot tip of the glow plug comes in contact with only a few fuel particles. The particles are vaporized there and ignite. This is how combustion is started. The helical heating filament of glow plugs is a binary filament consisting of a heating filament and a control filament. The helical heating filament rapidly heats the plug while the control filament, due to its positive temperature coefficient, acts as a current limiter thus preventing overheating.

Flame Starting System in the Intake Manifold

This system consists of flame plugs, solenoid valve, driving-starting switch, flame switch, flame-time control unit with power relay, indicator light and battery (Figure 62).

The flame plug is located in the intake manifold. The glow pin in the flame plug is surrounded by a vaporizing tube.

The tip of the glow pin in the flame plug is heated to about 1000 °C during pre-heating. The heat of the glow pin vaporizes the supplied fuel. This fuel vapor is mixed with the intake air. The high temperature of the glow pin is sufficient to ignite the air-fuel mixture thus creating a flame which heats the intake air.

The Control Unit

Glow-time Control Unit

Design

● Temperature-dependent timing element (evaluation electronics for the pre-heating time), Item ③ in the control unit.
● NTC resistor ④
● Timing element for the safety and short-circuit shutdown ②
● Short-circuit and overvoltage protection ⑤
● Power relay ⑥

Operation (Figure 63)

The sheathed-element glow plugs ⑦ are switched on via an internal power relay ⑥ by actuating the glow-plug and starter switch ①. A NTC resistor ④ measures the ambient temperature. The timing element ③ controls the pre-heating time as a function of the ambient temperature, as well as the safety cutoff time which is constant over the entire temperature range.

The engine is ready for starting when the start repeater lamp ⑧ goes out. Voltage continues to be applied to the sheathed-element glow plugs while the starter is being actuated so that the temperature required to initiate combustion is maintained. After starting has been completed, the current to the glow plugs is switched-off if post-heating is not required for a specific time. If the engine is not started after the start repeater lamp goes out, the glow-time control unit switches off the entire system after approximately 25 seconds (safety cutoff ②). Unnecessary heating of the glow plugs and battery discharge are avoided using this safety cutoff.

In the event of a short in the glow-plug circuit or overvoltage at the input terminals, the glow-time control unit switches off by means of the short-circuit cutoff ⑤ and can only be restarted after switching off the glow-plug and starter switch.

There are also other systems equipped with a plug monitor.

Flame-time Control Unit

The basic functions of this control unit are the same as those described for the glow-time control unit. However, the flame-time control unit can also drive a solenoid valve. The flame time control unit can also provide for an active flame both during and after starting.

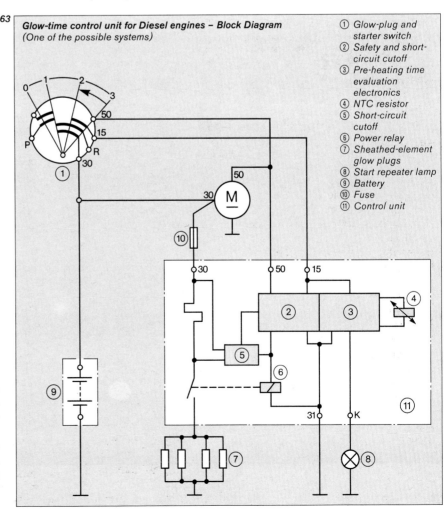

63 *Glow-time control unit for Diesel engines – Block Diagram*
(One of the possible systems)

① Glow-plug and starter switch
② Safety and short-circuit cutoff
③ Pre-heating time evaluation electronics
④ NTC resistor
⑤ Short-circuit cutoff
⑥ Power relay
⑦ Sheathed-element glow plugs
⑧ Start repeater lamp
⑨ Battery
⑩ Fuse
⑪ Control unit

As a supplement:

Automotive Electronics for Safety, Comfort and Reliability.

This publication describes how modern electronics is used to make motor vehicles safer and more reliable and how the operation of motor vehicles is improved and made easier. Safety systems offer increased protection for the driver and provide increased protection for the vehicle and its components. Electronics for comfort alleviate the load placed on the driver while driving.

The following are described:

● Antiskid system
● Airbag and seat-belt tightener
● Turn-signal flasher
● Turn-sequence control device
● Electronics used in the windshield cleaning system
● Overvoltage protection
● Car alarm
● Cruise control
● Trip computer
● Electronic transmission control
● Programmable power seats
● Automatic heating system
● Driver-guidance and data system

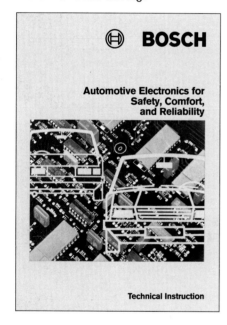

⊞ **BOSCH**

Automotive Electronics for Safety, Comfort, and Reliability

Technical Instruction

Technical Instruction
Concrete technical information covering the entire range of Bosch automotive products.